You, Me, and Technology:
Video Series Activities

YOU, ME, AND TECHNOLOGY: VIDEO SERIES ACTIVITIES

E. Joseph Piel
Minaruth Galey

Delmar Publishers Inc.®

NOTICE TO THE READER

Delmar Staff
   Executive Editor: Wes Coulter
   Associate Editor: Cynthia Haller
   Editing Manager: Barbara A. Christie
   Project Editor: Elizabeth A. Moslander
   Design Coordinator: Susan C. Mathews
   Publications Coordinator: Karen Seebald

For information address Delmar Publishers Inc.,
Two Computer Drive West, Box 15-015,
Albany, New York 12212

Contents

Preface

Now that our country has entered the Information Age, it is important for citizens to stop acting as though we were still in the Industrial Age. The *You, Me and Technology* video series is directed to helping you learn the new rules and the new ways of thinking necessary for you to succeed in this complex and technological era.

Many people have not yet learned these new ideas and do not realize what opportunities and occupations are available now, or that many more are coming in the future. As you complete the following activities with your class, you will be learning the skills and knowledge needed for your success. You will know more about today's problems, opportunities, and careers than most adults.

From our experience with other students, we expect that you will enjoy becoming involved in the real problems and situations in these activities that will help you become an effective citizen in this technological world.

E. J. Piel
M. Galey

PROGRAM 1

Living with Technology

By following a family through a single day's activities, we are introduced to the interaction of technology and people's lives. The changes in technology and its effect on people in the areas of food, clothing, communications, and transportation are demonstrated as we see clothing being made, news being delivered, people being transported, and food being processed and consumed. We are encouraged to look at both the advantages and problems which technology has brought us, and the trade-offs involved.

Activity 1 - 1
Comparison of Yesterday, Today, and Tomorrow

In the program "Living with Technology," the grandfather complains that radios are not as simple as they once were. He is told that with modern electronic devices it is often cheaper to buy a new one rather than to try to repair an old one. The concept "throw-away society" is introduced as a product of many advances in technology.

a. Some of the modern devices last longer than the ones they replaced. For example, compact discs last much longer than vinyl records, which lasted longer than the older 78 RPM records, which lasted longer than the Edison cylinders.

In the spaces list some throw-away items that are easier or less expensive to replace than to repair.

Clothing a sock with a hole in the toe; _____

_____

Toys _____

_____

Audio Equipment _____

_____

Other _____

_____

Now list some modern products which last longer than similar products made earlier.

Recording compact discs last longer than vinyl records that last longer than Edison cylinders;

_____

3

Tires _____

_____

Clothing _____

_____

Others _____

_____

b. A visit to the local radio/tv repair shop will provide a variety of burned-out vacuum tubes. You may find wiring diagrams of various circuits showing the increasing complexity of electronic devices. There will also be chances to check on the prices of chips, either at the store or in a catalog. Using an electronics store catalog, compare the price of a chip with that of a radio tube which is only one of the parts that make up a similar circuit.

_____

c. A personal computer is 1,000,000 times faster and contains more than 1,000 times as much space for memory as the original ENIAC (Electronic Numerical Integrator and Computer). The ENIAC had 18,000 vacuum tubes that needed a great deal of energy to operate and a complex system of fans and air conditioning to cool it off. The ENIAC needed 15,000 square feet of floor space (150′ × 100′) and weighed over 30 tons (60,000 pounds).

What size room would be needed to house enough ENIACS to replace one personal computer in memory space? _____

d. Radios which must be operated on house current have a warning statement telling us to disconnect the radio from the house electrical outlet before removing the protective rear panel. Battery-operated radios have no such warning. What are the risks associated with Walkman© and boom box type radios? What are the trade-offs of the various types of audio equipment? Use the following chart to record your answers.

TRADE-OFFS	
Type of Radio	
House Current Powered Radio versus Battery Powered Radio	
Small Portable Radio versus Stereo with Large Speakers	
Tape Recorder with Playback versus Compact Disc Player	

Activity 1-2
Technology and Consumerism

As technology is used in more facets of our lives, we become more dependent upon other people. We no longer grow all our own food, make our own clothes, or raise our own horses for transportation. We are consumers of someone else's efforts. Other people are consumers of our efforts. The successful persons are those who can convince more people to want what they produce.

One of the most effective methods for getting others to buy our product or service is to use television advertising. When we watch television on a commercial network we are bombarded by advertising.

To measure how much time is spent on advertising, you need to use the three Television Advertising charts, a watch that shows seconds, and a pencil. Watch three different programs: (a) a Saturday morning children's program, (2) a family-type situation comedy, and (3) a 30-minute news program. For each program, record information in the correct place on the chart as shown. Then answer the questions on the page following the charts.

TELEVISION ADVERTISING

Name		Date	Station	
Title of Program	**Time of Day**	**Code**	**Product Advertised**	**Length of Time**
	9:00:00	SI	Station ID	10 sec
	9:00:10	C	Toy	30 sec
	9:00:40			
	9:30:00	End		% Time
Totals		C		
		PRV		
		PROG		
		SI		
		PS		

Codes C—Commercial
       PROG—Actual program
       PS—Public service announcement
       PRV—Preview of another show
       SI—Station identification

TELEVISION ADVERTISING

Name		Date	Station	
Title of Program	**Time of Day**	**Code**	**Product Advertised**	**Length of Time**
	9:00:00	SI	Station ID	10 sec
	9:00:10	C	Toy	30 sec
	9:00:40			
	9:30:00	End		% Time
Totals		C		
		PRV		
		PROG		
		SI		
		PS		

Codes C—Commercial
        PROG—Actual program
        PS—Public service announcement
        PRV—Preview of another show
        SI—Station identification

TELEVISION ADVERTISING

Name		Date	Station	
Title of Program	**Time of Day**	**Code**	**Product Advertised**	**Length of Time**
	9:00:00	SI	Station ID	10 sec
	9:00:10	C	Toy	30 sec
	9:00:40			
	9:30:00	End		% Time
Totals		C		
		PRV		
		PROG		
		SI		
		PS		

Codes C—Commercial
       PROG—Actual program
       PS—Public service announcement
       PRV—Preview of another show
       SI—Station identification

JOHN DARLING

Fig. 1-1 Television Advertising! *(courtesy* The Star Ledger)

1. Does the commercial fit the consumer/viewer? _____ Explain.

   _____

   _____

2. What percent of time was spent on each coded item? For example, if 6 minutes of a 30 minute program is advertising, divide 6 by 30. This equals 0.2 or 20%. Add those percentage figures in the right margin of the chart.

3. Were the commercials accurate pictures of how the product actually works?

   _____ Explain in general terms. _____

   _____

   _____

4. How can consumers be sure that the products are accurately presented?

   _____

Activity 1-3
Changes in the Food Industry

Before the Industrial Revolution, most of the people in the United States lived on small family farms where they grew their own food and, for the most part, made their own clothing. The mechanization of agriculture made it possible for individual farmers to specialize in single crops, which they sold. They used the money to buy food, clothing, and other needed items. Today most Americans, even farmers, buy most of the food that their families eat. A number of different technologies are involved in bringing the food from the grower to the final consumer.

a. For the following list of foods, trace the food from the grower to the final consumer. Show the technologies involved in harvesting (mechanical picker), processing (freeze drying), transporting, storing, and delivering to the consumer.

	TECHNOLOGIES INVOLVED					
Food	Harvesting	Processing	Transporting	Storing	Advertising	Delivering
Milk	mechanical milker	cream separator	tanker trucks	electric refrigerator	TV ads	refrigerated trucks
Bread						
Meat						
Vegetables						
Fruits						
Fish						
Ice Cream						

b. One of the newer technologies for preserving food is by treating it with x-rays or radiation from radioactive materials. There are people who insist that this technology of irradiation is dangerous to the consumer. There are others who insist that there is no danger to the consumer and that the food can be stored longer, more economically, and in better condition than it can by any other process. Choose one side or the other of this argument. Look up the topic in the *Reader's Guide to Periodical Literature* in your library. Take notes from the articles you find, and be prepared to defend your opinion.

My opinion of food irradiation is:

_____

_____

_____

This is supported by: _____
(Name the source of your reference: Publication, Date of Article, Author); Statements in the

_____
article which support your argument.

PROGRAM 2

Decisions, Decisions, Decisions

Many people believe that only human beings can make decisions. They do not recognize that electrical and mechanical processors such as automatic oven controls, thermostats, and automatic door openers make decisions every day. Those machine processors are designed on a flowchart decision-making model that can improve human decisions, too. The sensor makes a measurement (temperature, time, pressure). The computer compares this with the desired situation and signals the processor to operate (turn off the oven, open the door).

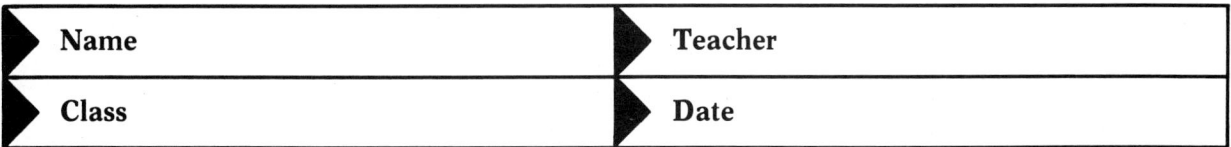
Activity 2-1
Feedback in the Home

The model for decision-making involving feedback from the program is labeled for the toilet tank system, Figure 2–1.

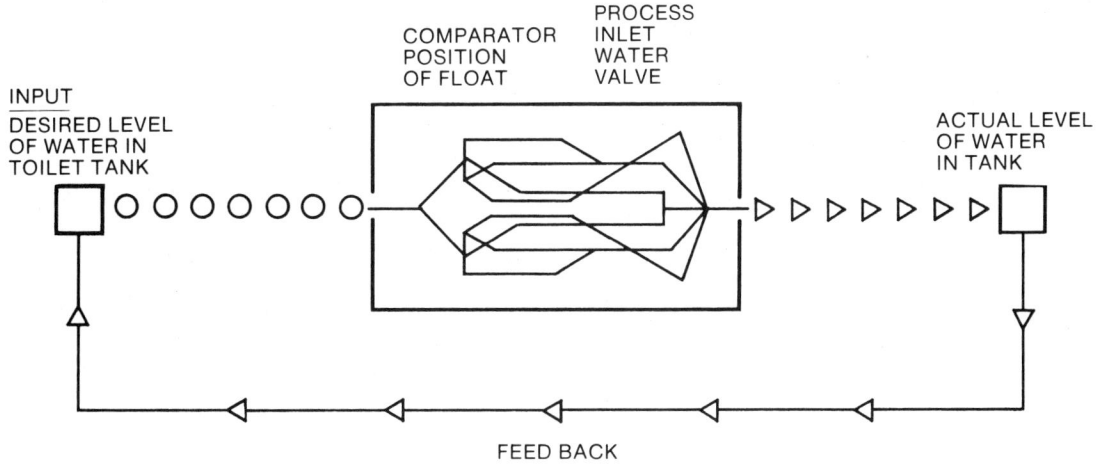

Fig. 2-1   Decision-Making Model

There are a number of automatic decision makers in the home depending on the number of automatic appliances. There are two which are part of almost every home in the United States: the thermostat and the control for filling the tank of the flush toilet.

As we look closely at the decision-making systems involving feedback, we find that we need to put more details in the basic diagram, Figure 2–2.

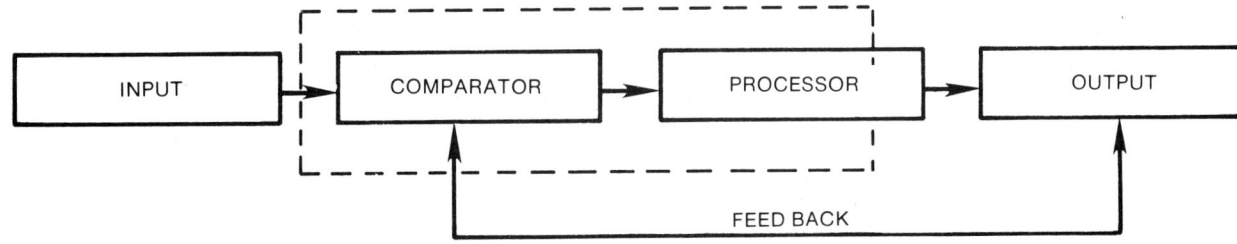

Fig. 2-2   A Feedback System

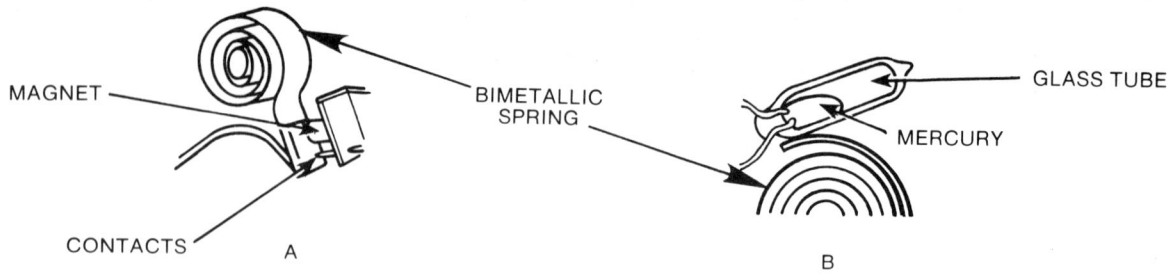

Fig. 2-3 Household Thermostats

In the case of the household heating system, the input is the desired temperature and the output is the actual temperature in the area of the thermostat.

Look at the diagrams of the thermostats in Figures 2–3a and 2–3b. Figure 2–3a shows an older thermostat with the open contacts. Figure 2–3b shows the new version in which the contact is made by a drop of mercury in a sealed tube making contact with the wires in the sealed tube.

Answer the following questions about how the heating system works.

1. The *input* is what you want. What is the input of the household thermostat?

   _____

2. The *comparator* compares what you want with what you get. What device

   does that? _____

3. The *processor* is the system which tries to give you what you asked for in the

   input. What part of the system does that? _____

4. The *output* is what comes out of the whole system. In the heating system, the

   output is _____

   Now explain how the heating system works. Use the terms input, comparator, processor, output, and feedback correctly in your answer.

   _____

   _____

   _____

   _____

   _____

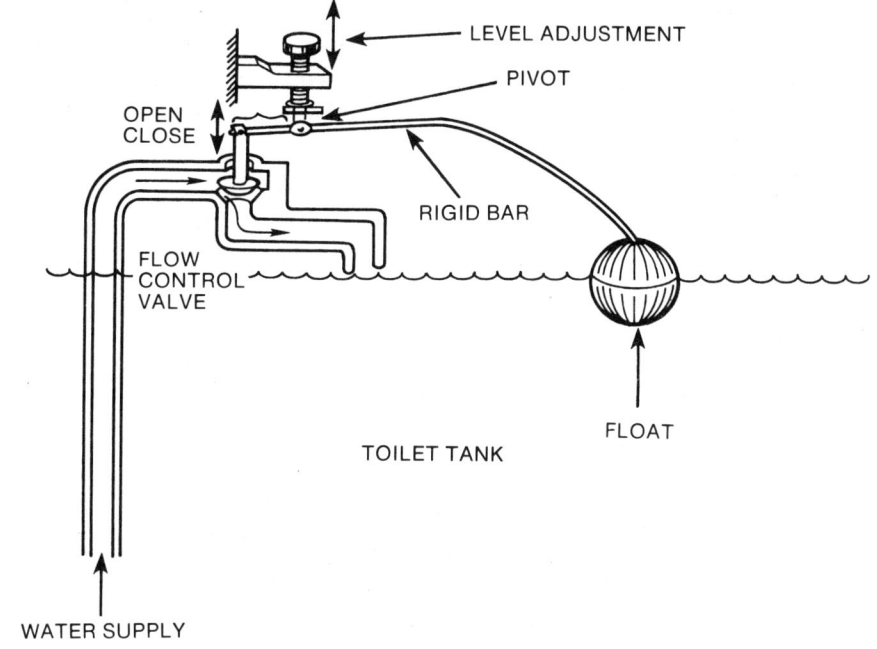

Fig. 2-4 Liquid Level Control for the Flush Toilet

Study Figure 2–4. Reproduce the basic feedback control flowchart from Figure 2–2. Label the flowchart using the names related to the flush toilet.

Write a description of how this system works as it allows water to fill the tank to a preset level.

_____

_____

_____

_____

_____

COMPARISON OF DECISION MAKERS IN THE HOME

	Input	Comparator	Processor	Output	Feedback
Household Heating System					
Toilet Tank					
Any Other Device					

Show how this system is similar to the house thermostat by filling in the following table. Be prepared to defend your answers.

Now that you have examined diagrams of thermostats and the operation of the water control system of your home toilet tank, you are ready to observe an actual thermostat in action.

The normal cooling system of the automobile sends water through a water jacket which surrounds the cylinders of the engine. The water picks up heat from the cylinders and travels to the radiator. Once there, it passes through very thin-walled pipes which make up the core of the radiator. Air rushing past this core cools the core and, therefore, cools the water. The water continues to circulate past the hot cylinders picking up heat and through the radiator losing some of the heat.

While it is necessary to take some of the heat from the engine, the liquid-cooled engine actually is more efficient if it runs at a temperature close to the boiling point of water. In order to maintain the optimum (the best answer when all factors are considered) temperature for the engine, we need a thermostat. Without a thermostat, the water might circulate so fast that the engine would not reach the optimum temperature.

The auto cooling system thermostat is a device that is placed in the cooling system in a spot where it will partially block the flow of water from the engine to the radiator until the optimum temperature is reached. At that temperature, the thermostat will not block the flow of water, Figure 2–5.

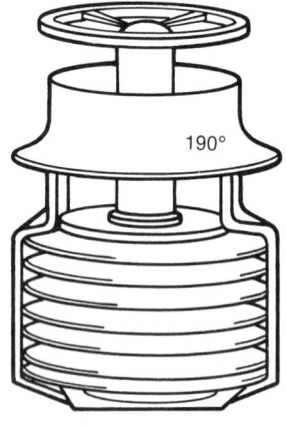

Fig. 2–5   Typical Auto Thermostat

Now you are to observe an auto thermostat in action. You will record both the temperature at which it opens to allow the water to flow freely and the temperature at which it closes to restrict the flow of water. Then you can make a flowchart of the feedback involved in the automobile cooling system.

MATERIALS (FOR THE SET-UP IN FIGURE 2-6)

- Auto thermostat
- Glass container in which to place the thermostat
- Thermometer that reads up to 100° C or 212° F
- Burner or hot plate to heat the water

PROCEDURE

1. Observe the thermostat. Look for the number that tells the temperature for which the thermostat is set. Some thermostats have two numbers; others, none.
2. Testing the thermostat. Using a string tied to a support, hang the thermostat as shown in Figure 2–6.
3. Apply heat to the beaker until the thermostat opens. Refer to Figure 2–5 to determine when the thermostat has opened.
4. Record the temperature at which it begins to open. _____
5. Record the temperature at which it is fully open. _____
6. Allow the water in the beaker to cool. Explain why you DO NOT ADD COLD

   WATER to the beakers. _____

   _____

   _____

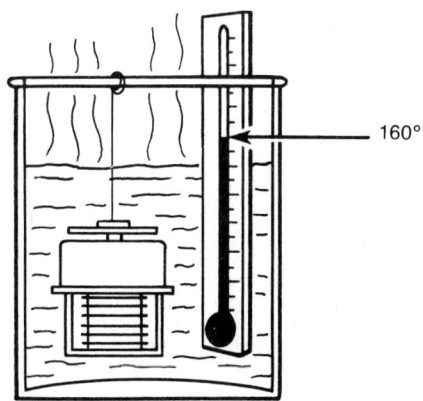

160°

Fig. 2-6   Equipment Set-up for Testing the Thermostat

7. Record the highest temperature at which the thermostat begins to close.

_____

8. Record the temperature at which it is fully closed. _____

QUESTIONS

1. If there was a temperature indicated on the thermostat, compare that number with any of the four numbers that you recorded. With which temperature did the number stamped on the thermostat agree? _____

2. What is the range of temperatures for the thermostat you tested? (The range is the difference between the temperature at which the thermostat opened and the temperature at which it closed.) Opened _____ Closed _____
   Range = _____

3. How does this thermostat affect the amount of heat that you get from the heater in the passenger section of the car?

_____

_____

4. How can you tell that the thermostat is not working, just by sitting in the car?

_____

_____

_____

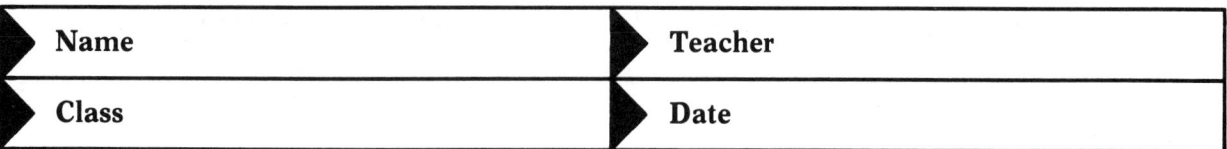
Activity 2-2
Delay in the Feedback Circuit

You are taking a shower. When you get in the shower, you adjust the hot and cold faucets (or the single mixer faucet) until you get the right temperature. Someone in the kitchen turns on the hot water. After a short delay (depending on the distance between the hot water tank and the shower), the water gets cold! You turn up the hot water—no change. You turn it up further. Suddenly it is too hot! You turn it down—no change. Suddenly it is too cold! The system is becoming unstable. If the delay is really long, the problem becomes serious enough that you turn off the shower and start all over again.

The following activity will show how delay in the feedback loop causes a system to become unstable.

PROCEDURE

1. With a group of 3 students, trace the line in Figure 2–7 while looking at it. Now trace the line while blindfolded. Move your hand from left to right at a constant speed as your partner tells you to move your pencil up and down. Remove the blindfold and notice the difference between the two tracings.

2. Now arrange to have a third person watch you tracing while you are again blindfolded. The third person will point up or down to your first partner who will then tell you to move your pencil up or down as you continue to move your hand from left to right. Notice the difference between this tracing and the original. Switch jobs so that each partner has a chance to trace. Compare

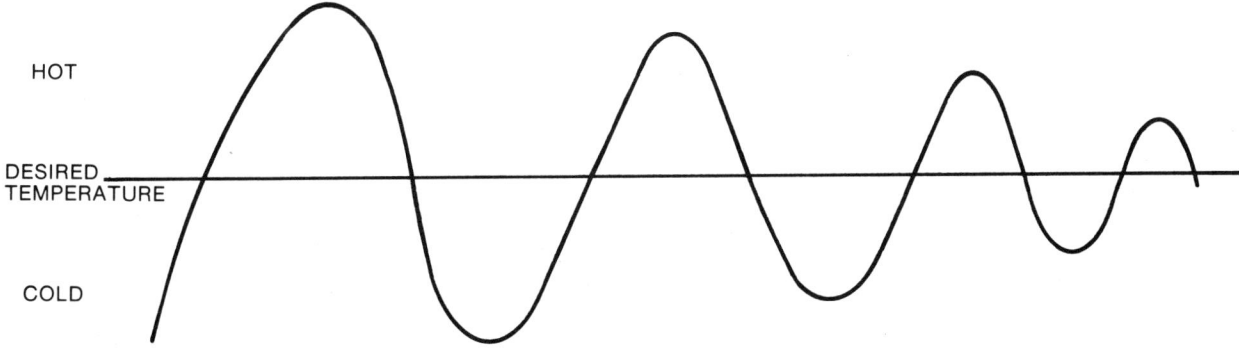

Fig. 2-7   Delay and Feedback

21

these tracings with the description of the shower story. Assume that up is hot, down is cold, and the straight line is the desired temperature.

Compared to the shower:

a. The first tracing is similar to _____

_____

b. The second and third tracings are similar to _____

_____

Describe the problems that may result from delayed feedback in the following situations:

1. Driving a car

2. Receiving results of a physical examination

3. Political polls

4. Household thermostat

5. Toilet tank level control

6. Aircraft instrument landing system

7. Publishing the results of a nuclear accident

Activity 2-3
Technology Assessment

In the program, Russell Train, then head of the Environmental Protection Agency, said that we should try to assess the effect of a proposed technology before the technology is introduced. Richard Garwin did that in the case of the SST. He predicted that it would not be economically or environmentally desirable for the United States to build a supersonic transport.

Use the chart on page 26 to do a technology assessment of genetic engineering; that is, changing the genes of animals or plants. The left-hand column lists a potential technology, and the next column gives some possible results of the introduction of the technology. There are also blank spaces for you to list any other possible results. The next three columns give you a chance to say how likely you think it is that the suggested results might follow. The last three columns give you a chance to say how favorable you think the result would be.

PROCEDURE

1. Complete the form suggesting other results for D, E, and F.
2. Meet with three other students to discuss and agree where the checks should be in columns 3 through 8.
3. Discuss your group decision with the rest of the class. Be prepared to defend your decisions.
4. With the class, decide in an oral discussion whether the technology should be encouraged or stopped. Then describe ways this technology should be encouraged or stopped.

ANALYZING A TECHNOLOGY—SOCIETAL SITUATIONS

Development or Situation (1)	2. Might Result In	How Likely Is It That the Result Will Follow the Development or Situation?			What Will the Effect of the Result Be?		
		3. Virtually Certain	4. Possible	5. Almost Impossible	6. Favorable	7. Of Little Importance	8. Detrimental
Genetic engineering to allow parents to control how tall their children would grow.	A. Parents choosing to grow tall sons up to 7'10"						
	B. Parents choosing to have daughters no taller than 5'5"						
	C. Government control of the technology to produce people of various sizes for specific jobs.						
	D.						
	E.						
	F.						

PROGRAM 3

The Technology Spiral

Technology is basic to the exciting history of human beings who have advanced from living in small family groups to living in great cities. The change from simple hand tools to mass production and automation has produced a wealth of material pleasures as well as new kinds of limits. All of these changes have greatly increased the demands for energy and resources that have produced critical shortages today.

Activity 3-1
Tools and Revolutions

The *Technology Spiral* describes world development in terms of four eras: the Toolmaking Revolution, the Agricultural Revolution, the Industrial Revolution, and the Information Revolution.

PROCEDURE

1. Describe a tool (device or machine) that is important to each Revolution.

   a. Toolmaking _____

   b. Agriculture _____

   c. Industrial _____

   d. Information _____

2. Form groups of four people. Have students in your small group compare descriptions. Choose the best description and present it to the whole class. Take notes to be prepared to defend your choice in terms of the criteria you set for the best tool.

   _____

   _____

   _____

Activity 3-2
Matching the Tool
to the User and to the Job

Tools work best when they give the best match between the human user and the job to be done. One of the tools that has been accepted by individuals and businesses for presenting the written word is the typewriter. To be specific, it is the typewriter keyboard. Typewriters, word processors, and computers all use keyboards that have the letters in the same relative positions. Is this the best match between the human user and the job to be done? To answer this question, let's look at (1) the job to be done, (2) the human characteristics of the user, and (3) the keyboard.

THE JOB TO BE DONE

The job is to present written documents (letters, reports, books) in readable form in the most efficient manner. "Efficient" here means the shortest time with the fewest errors.

If the keyboard writes one letter at a time, it is valuable to know which letters are used more often and which are used least often. We should then compare the letters to the human user to determine which fingers are easiest to use when typing. We should then arrange the letters on the keyboard to match the ones used most often with the fingers which are easiest for the typist to use.

What is the rate at which various letters are used in writing English? For standard English, we find that in 1,000 letters, each letter will appear about the following number of times:

E 132 _____	S 61 _____	U 24 _____	K 4 _____
T 104 _____	H 53 _____	G 20 _____	X 1 _____
A 82 _____	D 38 _____	Y 20 _____	J 1 _____
O 80 _____	L 34 _____	P 20 _____	Q 1 _____
N 71 _____	F 29 _____	W 19 _____	Z 1 _____
R 68 _____	C 27 _____	B 14 _____	
I 63 _____	M 25 _____	V 9 _____	

Starting with the sentence following the heading "Activity 2" that begins "Tools work best..." and ending with "...for the typist to use," there are about 1,000 letters. If each student in a class of 26 is assigned a letter and is asked to count the number of times that letter is used, you will know how close that paragraph is to the average in English letter usage. In the space at the right of the letter, list the number of times each letter is represented.

From your count, and from the average presented in the activity, list the 10 most often used letters and the 20 least used.

Most Often Used		Least Often Used	
1 _____	6 _____	17 _____	22 _____
2 _____	7 _____	18 _____	23 _____
3 _____	8 _____	19 _____	24 _____
4 _____	9 _____	20 _____	25 _____
5 _____	10 _____	21 _____	26 _____

HUMAN CHARACTERISTICS

Using Figure 3–1, place your fingers on the shaded keys. Move your fingers to each of the other keys pictured on the keyboard.

1. Which ten numbered keys were easiest to reach and press?

_____ _____ _____ _____ _____ _____ _____ _____ _____ _____

2. Which ten numbered keys were most difficult to reach and press?

_____ _____ _____ _____ _____ _____ _____ _____ _____

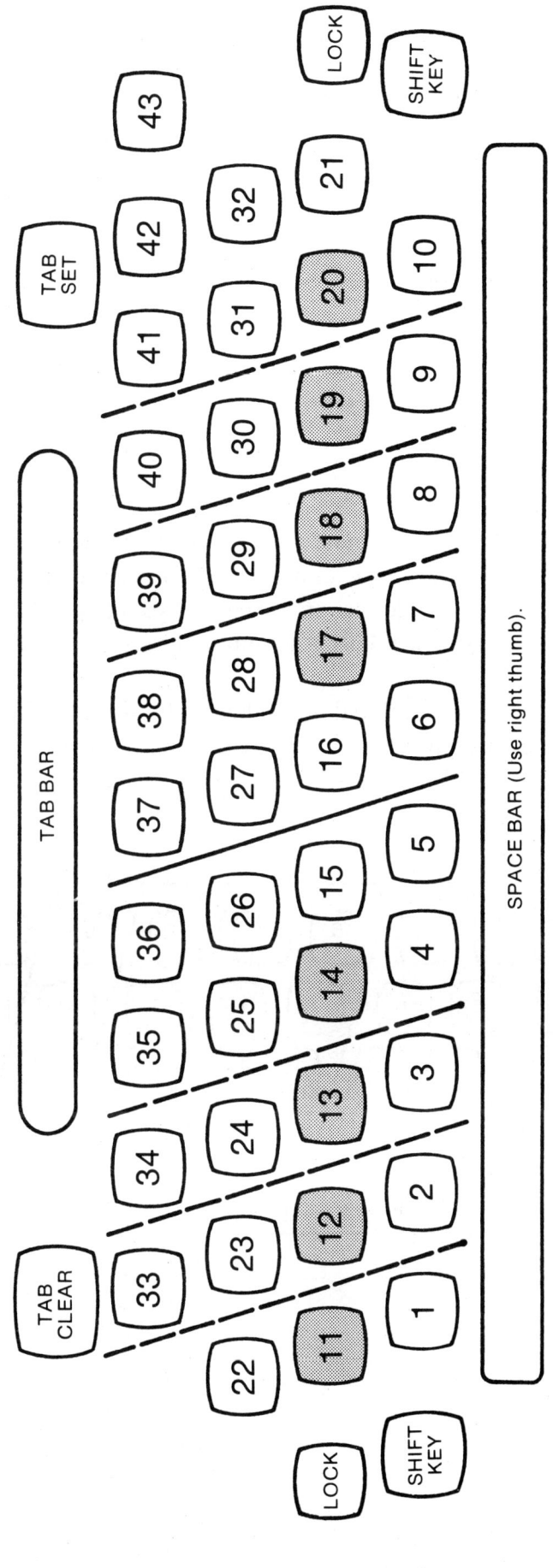

Fig. 3-1 Numbered Keyboard

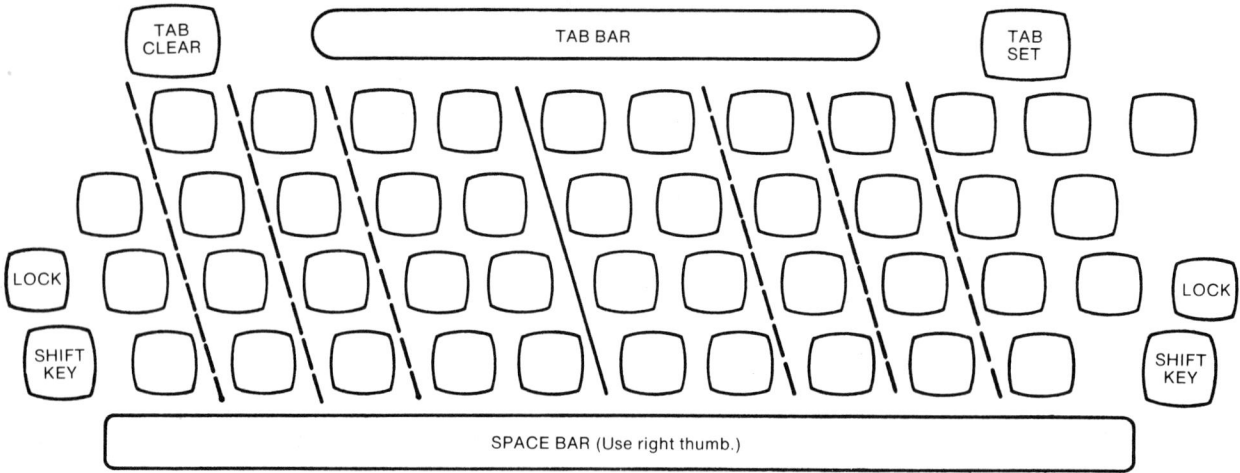

Fig. 3-2   Blank Keyboard

MATCHING THE USER TO THE JOB

Based on what you found to be the most and least often used letters, and which keys are easiest and hardest to press, you are now in a position to design your own keyboard. Using Figure 3–2, write the most often used letters on the easiest keys to press, and put the least often used letters on the hardest keys to press. Fill in the remaining six letters and the ten digits 0—9 on the appropriate keys.

Now, compare the keyboard that you designed with the standard keyboard in Figure 3–3. While the keyboard that you designed might not be the best for all people, it does seem easier to use than the standard keyboard. How could the designer of the standard keyboard be so far off?

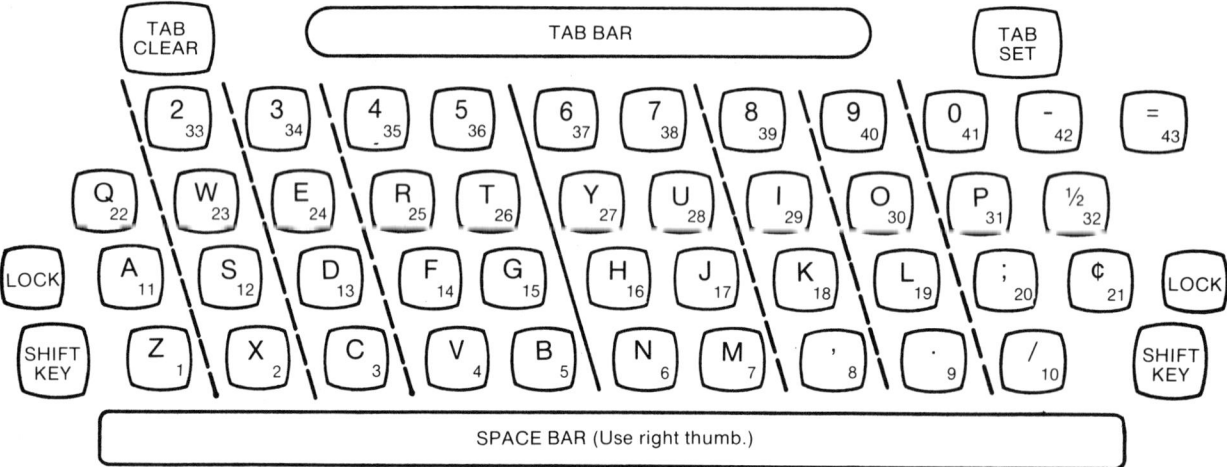

Fig. 3-3   Standard Keyboard

Fig. 3-4   Early Model Typewriter

The designer had a constraint that we did not include in our decision-making task. That constraint was to keep the typewriter keys from jamming together when the typist typed fast. On the early typewriter, the keys were operated through a series of lever arrangments.

Electronic typewriters or those which operate with a ball have eliminated all of those lever

problems, but we still have the old keyboard. Why? _____

_____

_____

What steps do you think need to be taken in order to move from the standard keyboard to one

which would allow for faster typing? _____

_____

_____

_____

ACTIVITY 3-3 TECHNOLOGICAL REVOLUTIONS AND JOBS

When automatic dialing was introduced in the telephone system, people were concerned that this would result in much unemployment among the telephone switchboard operators. These operators connected the caller to the person being called by plugging wires into a switchboard. It has been calculated that if we did not have automatic dialing and still had the same number of telephones that we have today, about 100,000,000 switchboard operators would be needed to do the job. It would also take an average of 5 minutes from the time you picked up the telephone until you were connected to any telephone outside your calling area.

Check with your local telephone company to find out how many operators are employed now compared with the number 30 years ago. How many more telephones are there now compared to 30 years ago? _____

_____

Describe one job in agriculture that has been eliminated by the use of machines. Explain the

advantages that have resulted from the use of that machine. _____

_____

_____

Describe one job in industry that has been eliminated by the use of machines. Explain the

advantages that have resulted through the use of that machine. _____

_____

_____

37

Describe one job in industry that has been created by the use of new technologies. Explain the advantage of that job over the job that was eliminated. _____

_____

_____

What "Information Age" machines are available to you in school that were not available to your parents? _____

_____

_____

What machines were available to your parents, but not your grandparents? _____

_____

PROGRAM 4

Energy and Societies

When human beings learned to use fossil fuels, they reduced much of their work. With large supplies of iron and coal, steel was made in great amounts. Engines built of steel greatly increased the productivity of the industrial nations. The use of energy increased so much that mining and burning fossil fuels were destroying our environment.

The need for alternatives to fossil fuels produced new technologies. These new technologies are being developed to give us at least 5 possible new sources of energy. We must take care in discovering and evaluating the trade-offs in the use of these fuels and the damage to the environment.

Activity 4-1
What Is the Appropriate Technology?

As we look at the issue of energy conservation, we are often faced with a problem. One such problem involves the size of the energy supplies with respect to the job to be done. For example, it is important that we choose a home heating system with the optimum output. Optimum means the most efficient output for the location, size, and characteristics of the house. A furnace that is too small for the job will run for too long a period of time, especially on very cold days. A furnace that is too big will burn too much fuel especially on days when not much heat is needed. A similar issue can be examined regarding the most efficient method for heating water in the home. One method is the one-cup immersion heater. It consists of a small coiled heating element that is submerged in a cup of water and then plugged into an electrical outlet.

The usual method of using electricity to heat slightly larger amounts of water is to put the water in a pan on top of a hot plate. Which is the more efficient system? We would probably agree that for heating a gallon of water, the hot plate system is best. For heating one cup of water, the immersion heater is best. But which is best for two cups, or two quarts? In this experiment you will try to find out the quantity of water at which the hot plate and immersion heater are equally efficient. You will then know which is the best technology for any amount of water to be heated.

In the *Power Requirements* table which follows, record the power needs of the immersion heater and the hot plate. To find the energy used, multiply the power in watts by the time in seconds; that is, Energy = Power × Time. For example, a 200-watt device operating for 90 seconds = 200 watts × 90 seconds = 18000 watt-seconds. You may use fluid ounces or cubic centimeters for your units of liquid measure, depending on the measuring instruments available.

PROCEDURE

Start with 8 ounces of water if using the English system or 250 cc if using the metric system. Increase by 2 oz or by 50 cc (depending on your system) for each additional trial. In order for the data to be accurate, you should start each trial with the immersion heater and the hot plate at room temperature. This will be easy for the immersion heater, but a bit more difficult for the hot plate. How can you organize for this to happen? _____

_____

What other difficulties will you need to overcome? _____

_____

_____

POWER REQUIREMENTS

Quantity of Water	Immersion Heater Power _____ watts		Hot Plate Power _____ watts	
oz or cc	Time in seconds	Energy in watt seconds	Time in seconds	Energy in watt seconds

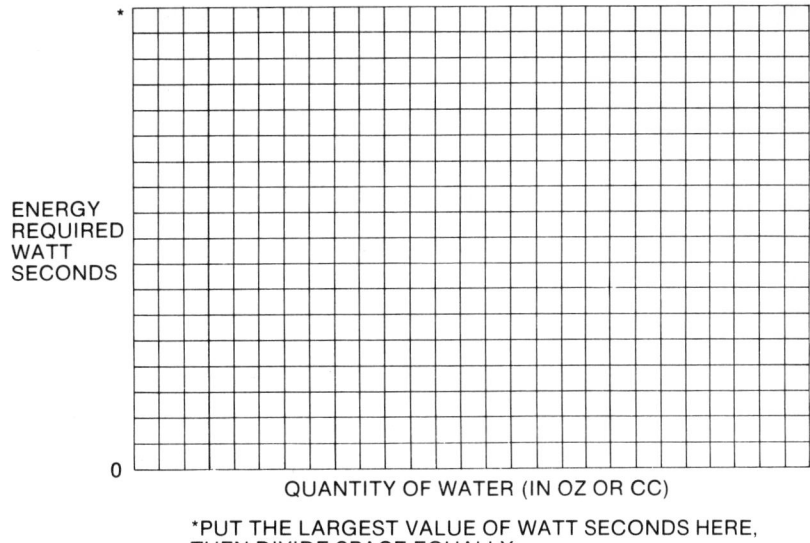

ENERGY
REQUIRED
WATT
SECONDS

0

QUANTITY OF WATER (IN OZ OR CC)

*PUT THE LARGEST VALUE OF WATT SECONDS HERE,
THEN DIVIDE SPACE EQUALLY.

**Fig. 4-1   Comparison of Energy Required to Heat Varying Amounts of Water
Using Immersion Heater and for Hot Plate**

After you have completed enough trials (about 8) so that you feel that you have enough data to make the decision on which system you will use for various amounts of water, record your results on the graph in Figure 4-1. Use a solid line for the hot plate and a dashed line for the immersion heater.

QUESTIONS

1. Do the lines representing the type of heater ever cross? _____

2. If they do, at what quantity of water did they cross? _____

3. How is this experiment similar to the furnace situation?

_____

_____

_____

_____

4. How is this experiment different from the furnace situation?

_____

_____

_____

_____

5. What other energy problems might be solved by this type of experiment?

_____

_____

_____

_____

Activity 4-2
Energy Trade-offs

As we study the energy situation in the world we find that there are a number of problems. The sources of energy differ in: (1) where they come from, (2) how they are used, and (3) the effects of their use on the environment.

The *Energy Problems* table lists a number of sources of energy in the left-hand column and a list of problems across the top. This makes a set of boxes, called a *matrix*. Your teacher will assign your group to consider one of the sources of energy. In the matrix put a check in each box in which your source of energy is responsible for a problem. Be ready to explain what that problem is. For example, water power is not available where the ground is flat, so you would put a check in the location column opposite water power.

ENERGY PROBLEMS

Source of Energy	Location	Cost	Eventual Shortage	Danger In Use	Affects Climate	Causes Pollution
1. Oil						
2. Gas						
3. Coal						
4. Solar						
5. Wood						
6. Water Power						
7. Wind Power						
8. Geothermal						
9. Nuclear Fission						
10.						

When each of the groups have agreed which boxes should have check marks, the teacher will combine all of the checks in a large chart on the chalkboard or on a transparency. After discussing the reasons for the placing of the checks, complete your matrix by putting checks in all of the appropriate boxes.

Meet in your group again. Discuss the ways in which you would recommend that the trade-offs could result in the least number of problems. Don't just trade off one source against another. Remember that the tenth row was left empty. That row could have been labeled conservation, recycling, burning garbage, or any other source that you think would be better than the one you or anyone else were assigned. After discussion with your group, explain what the trade-off is between the new tenth source and the one you were assigned.

Trade-offs:

(Last Row) Your suggestion   vs.   Your assigned energy source

_____

_____

_____

_____

_____

_____

_____

_____

Activity 4-3
Solar Convection Heater

Many so-called energy saving devices and systems require an outside source of energy to operate. For example, solar hot water systems for home heating require electricity to operate the fans and pumps.

In this activity, you will build a one-room solar heater which needs no energy source other than the sun to operate. The solar heater is placed outside the room and extends into the room through a partially opened window. The solar heater assembly looks similar to the diagram in Figure 4–2.

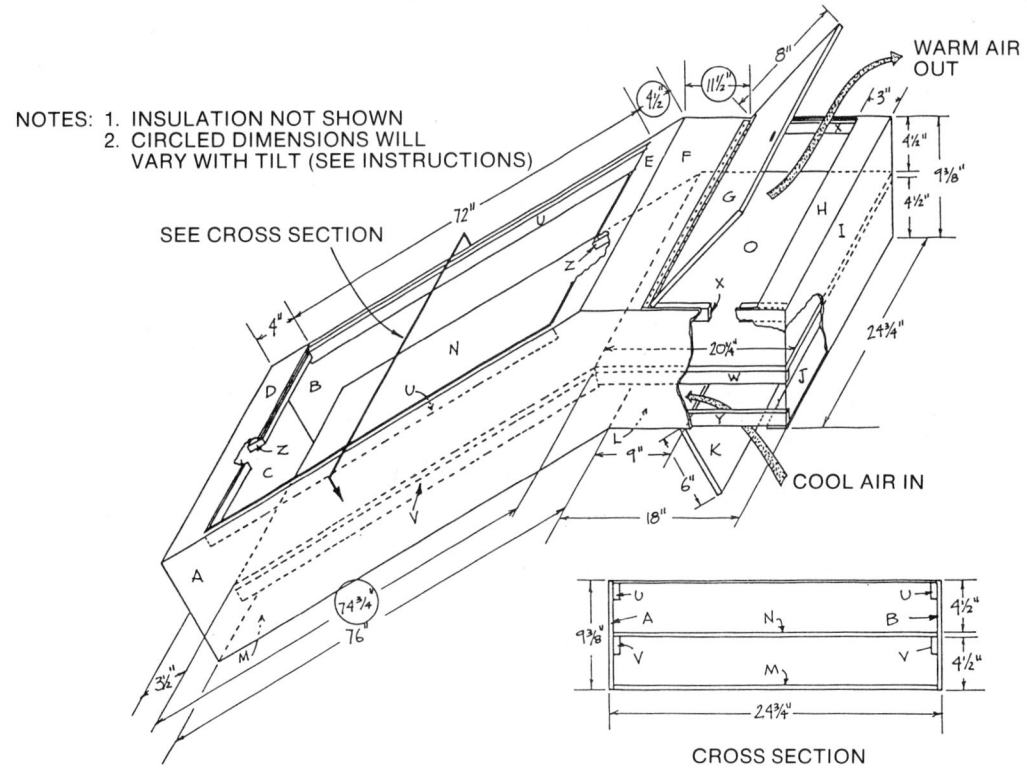

NOTES: 1. INSULATION NOT SHOWN
2. CIRCLED DIMENSIONS WILL VARY WITH TILT (SEE INSTRUCTIONS)

SEE CROSS SECTION

WARM AIR OUT

COOL AIR IN

CROSS SECTION

Fig. 4-2   Solar Window Collector

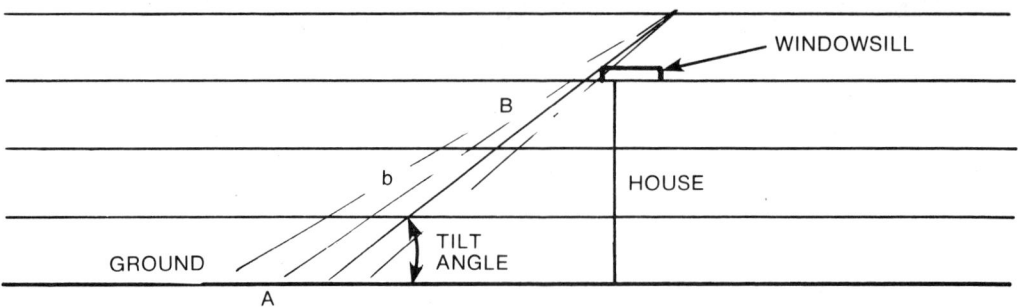

Fig. 4-3 Tilt Angle

This heater is best for a room that faces within 15 degrees of south. This is 165–195 degrees measured clockwise from north. The *angle of tilt* will depend on your angle of distance from the equator. The angle of tilt is determined by adding 10 degrees to your latitude. For example, the latitude of Albuquerque, New Mexico is 35°. Therefore, the correct angle of tilt is 45° (or 35° plus 10°). The latitude of Boston is 42°; Spokane, Washington, 48°; and Washington, DC, 39°.

Once you have determined the tilt angle, you can decide on the length of the bottom slanted surface of your collector by using a stick and a protractor. Put the stick on the windowsill with the end on the ground. Move the end of the stick toward or away from the house until the tilt angle is the one you have chosen, as in Figure 4–3. The distance from A to B determines the length of the bottom section of your collector (distance b of Figure 4–3).

You are now ready to build the passive solar collector.

MATERIALS (COST IS ABOUT $60)

- 2 sheets of 4' × 8' exterior grade plywood ⅜" thick
- 5 pieces of 1" × 2" furring strips, 8' lengths
- 2 pieces of piano hinge, 2' lengths
- 2 lb of 4d finishing nails
- 1 sheet of plastic glazing, thickness 0.10 nominal (24" × 72")
- 40 sq ft styrofoam insulation, 1" thick
- 2 tubes of caulking (silicon rubber is best)
- 1 bottle of carpenter's glue (large)
- ½ gal exterior paint primer (latex)
- ½ gal exterior paint (latex)
- 1 qt flat black exterior paint
- 2 magnetic cabinet latches
- 2 sheets of sandpaper (medium)
- 2 rolls of double-faced tape

TOOLS

- 1 saw, circular, saber, hand
- 1 hammer
- 1 utility knife
- 1 caulking gun
- 1 protractor
- 1 pencil
- 1 straightedge or meter stick
- 1 tape ruler
- 2 2"–3" paint brushes (throw away quality)
- supports for wood cutting
- 1 carpenter's snap line—optional

PROCEDURE

The dimensions given are for a location at 41° N latitude (angle of tilt = 51°). If your location differs from this by more than 5°, you may wish to alter dimensions to conform to your own tilt angle. General instructions are given here.

1. Lay out the pattern for the pieces on the two sheets of plywood using the dimensions shown in Figures 4–4 and 4–5. Letter each part as shown in these diagrams. (*Note:* If you are using a different angle of tilt, the dimensions of Parts E, F, N, and O will change as will Angle S. To find these dimensions, do steps a through k first.) In order to help you, Figure 4–5 is labeled in the same order as the following steps.

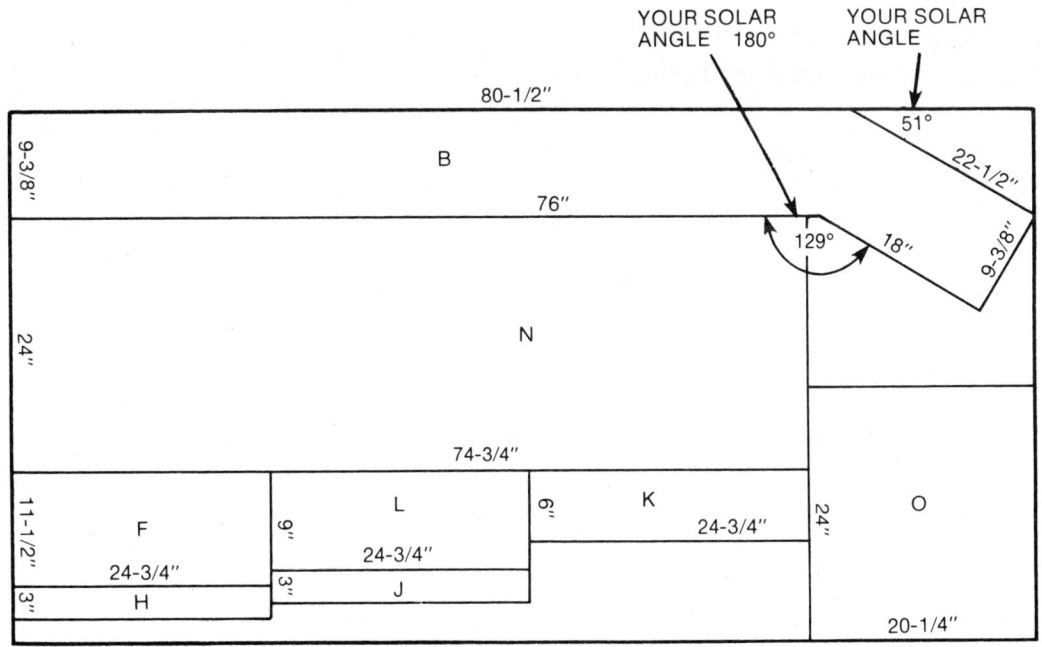

Fig. 4-4   4′ × 8′ Sheet of ⅜″ Exterior Grade Plywood

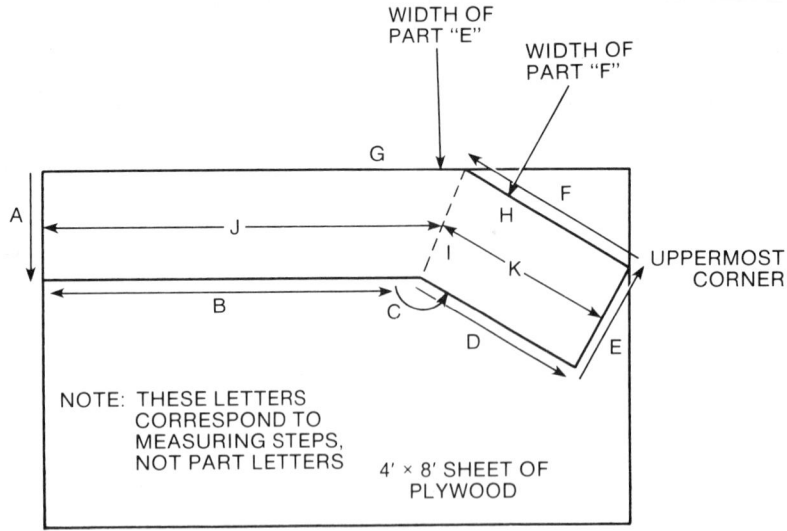

Fig. 4–5   Steps in Building a Solar Window Heater

Starting from the lower left-hand corner of one of the plywood sheets:

a. Measure and mark off 9⅜" along the bottom.

b. Draw a line 76" long parallel to the left-hand side.

c. Using the protractor and the end of this line as the start, measure and mark off Angle S so that it equals 180° minus the tilt angle.

d. Draw a line 18" long at the Angle S you have just measured.

e. At the end of this line and at right angles, draw a line 9⅜" long.

f. Now at the end of the line drawn in *e* and at an angle of 90° to it draw a line extending to the left-hand side of the plywood. You have just drawn the pattern for Part A which can be copied onto the second sheet of plywood to make Part B. You are, however, not finished.

g. Go back to the lower left-hand corner of the plywood. Measure up a distance of 76" along the left side. Whatever is left of Part A along this edge is the width of Part E.

h. Now move to the uppermost corner of Part A. Measure off a distance 11" along the line you drew in step *f*. Whatever is left of this line is the width of Part F.

i. Draw a line connecting the two bends in Part A as in Figure 4–5. Mark the midpoint of this line.

j. Mark off a distance of 1¼" down from the bend of Part A. From this point to the bottom of the plywood is the length of Part N (76¾").

k. The distance from the upper 9⅜" edge to the midpoint found in step *i* is the length of Part O.

Having thus found the dimensions of Parts A, B, E, F, N, and O, mark them off on the plywood as in Figures 4–4 and 4–5. Now you may continue with the construction procedure.

2. Cut out the parts.

3. Sand the rough edges.

4. The furring strips act as supports and should be cut in the following lengths. (Those with * will differ based on your dimensions.)

   2 pieces 72″ long (labeled U in Figure 4–2)
   *2 pieces 74¾″ long (V)
   *2 pieces 20¼″ long—these will later be cut somewhat shorter to fit (W)
   2 pieces 8″ long (X)
   2 pieces 6″ long (Y)
   2 pieces 22½″ long (Z)

5. Glue and nail Parts V through Y onto Parts A and B as the plans show.

6. Center Part Z on Parts D and E. Allow Part Z to extend ¾″ past the edge. Glue and nail Part Z to Part D and E. **Caution:** Since the wider face of each furring strip is being nailed, the nails will project out after nailing.

7. With the exception of Parts N and O (the absorber and divider), glue and nail the box together as plans show, including the hinges of Parts G and K. Attach magnetic catches to Parts G and K.

8. With the exception of the upper halves of Sides A and B, install the insulation using double-faced tape to hold in place. The following surfaces do not and will not receive insulation: G through K.

9. Insulate the bottom surfaces of Parts N and O. Remember to leave 1″ at each edge so that these parts will fit onto the furring strips.

10. Prime and paint the outside of the box.

11. Prime and paint the upper surface of Parts N and O with flat black paint.

12. Glue and nail Parts N and O into the box.

13. Caulk edge where Parts N and O touch each other so that there will be less leakage from the bottom chamber to the top.

14. Now insulate the upper halves of Sides A and B.

15. Lay the plastic glazing onto the furring strip supports. (The edges of the glazing may have to be sanded to fit.)

16. Apply a thick bead of caulking around the entire edge of the glazing and on the edge where Surfaces E and F meet.

Congratulations! You have finished. Place the solar convection heater into the window you have chosen. Use the scrap wood and some tape to block any openings. Check the airflow through the collector by watching smoke or dust as it flows from the room through the bottom opening and into the room through the top opening. Check the temperature at both the bottom and the top on cloudy and sunny days.

PROGRAM 5

Health and Technology

The first successes in curing and then preventing some illnesses were outstanding. Now we have successes in life-saving and life-prolonging technologies as well. Science has produced new knowledge, and technology has applied that knowledge for the techniques of organ transplants and even the use of artificial organs as transplants. However, this progress in science and technology has made it necessary to carefully consider the trade-offs among costs, risks, and benefits.

Activity 5-1
Whose Life Will Be Saved?

One part of the program deals with technological systems for saving lives. When there are fewer of these systems than there are people who need them, the question is asked "Whose life will be saved?"

In some hospitals, the decision as to who will get the next available kidney is based on a first-come basis. The person who has been on the waiting list longest, and whose tissue type matches that of the donor, gets the kidney. In other hospitals, a committee decides who gets the next kidney. Who do you think should be on the committee in terms of background and present position?

1. _____

2. _____

3. _____

4. _____

In your committee of four students decide on what basis your group will make your decisions:

1. _____

2. _____

3. _____

4. _____

5. _____

The following five people are all of the same tissue type as the donor. Rate them based on your criteria, and decide who should get the kidney. Note: Each person is equally ill and will probably die within a year if a new kidney is not received.

Joe

Sixteen years old, B-average student in High School. Father died of kidney failure. Plans to become a teacher.

Mary

Thirty year old mother of one ten year old child. Active in church work. Plans to go back to work as a laboratory technician when child enters junior high school.

Bill

Forty-five year old father of two children who are in college. Has been missing appointments for dialysis treatments. Says he cannot stand to be connected to a machine.

Dr. Sam

Thirty-five year old neurosurgeon, unmarried but engaged. Deals with head injuries of people involved in auto accidents.

Betty

Twenty-nine year old mother of two children living on welfare. Active in welfare rights of tenants. Attending job training program. Expects to get a job as medical technician.

Did your committee's choice agree with that of the other committees in the class? _____

What were the differences? _____

_____

What were the concerns you and your group had in making these decisions? _____

_____

_____

_____

_____

_____

_____

_____

_____

_____

Activity 5-2
Using Our Senses

As we lose the use of one of our senses we often develop greater sensitivity in another. For example, blind people need to sharpen their senses of hearing and touch in order to receive as much communication as possible. How well do your senses of hearing and touch enable you to receive information?

PROCEDURE

Your teacher will give your group a box with a number of different objects in it. You may tip it, weigh it, and shake it gently. Listen carefully while you are doing that. Try to coordinate what you hear with what you feel. Are the objects round, flat, cubical, or irregular? Are any of them fastened together? Are they fastened to the box? Record your answers in the data table that follows.

WHAT'S IN THE BOX?

(Please use check marks for your answers)

My Judgment:

_____ Round _____ Flat _____ Cubic _____ Irregular

_____ Fastened together _____ Free to move _____ Fastened to the box

My Partners' Judgments:

_____ Round _____ Flat _____ Cubic _____ Irregular

_____ Fastened together _____ Free to move _____ Fastened to the box

---

How I came to my conclusions. _____

_____

When you and your partners have finished ask for permission to look in the box. How good was your hearing and touch compared to your eyesight?

_____

_____

When you are blindfolded, your teacher will give you some objects to feel. Describe the objects and have your partner write what you say.

1. _____

_____

2. _____

_____

3. _____

_____

What were the objects which you could see when you took off the blindfold?

1. _____

2. _____

3. _____

Activity 5-3
When Do We Stop
Using Technology?

With the advances in medical technologies it is often possible to keep people alive even though they will never regain consciousness.

Example:

Bill, a high school senior, has been injured in an automobile accident. His head injuries are so severe that he will never regain consciousness. However, his heart and lungs have not been severely damaged.

The cost of keeping Bill alive will drain his family's bank account, and his brother and sister will not be able to go to college. His parents cannot agree whether to keep him alive or to let him die. The hospital administration points out that the intensive care unit of the hospital is crowded and every possible space is needed for present and future emergency cases.

The doctor in charge has the following options:

1. Do everything technically possible to keep Bill alive.
2. Discontinue intravenous feeding and let Bill die.
3. Discontinue use of the other technologies which are keeping Bill alive.
4. Give Bill an overdose of one of the drugs now being administered.

PROCEDURE

Divide into groups of 4 students. Discuss and answer the following questions:

1. Which option should be taken? _____

_____

2. What influenced your decision (ethics, economics, religion, legal issues)?

_____

_____

_____

3. Who should really have the responsibility for making the decision? Why?

_____

_____

_____

(This activity was adapted from materials of the National Center for Bioethics.)

PROGRAM 6

Feeding the World

In most of the United States, food is so abundant that it is regarded more as a source of pleasure than as a necessity for life. The stories of starving colonists contrast sharply with the great variety and quantity of food in the supermarkets of this nation today. With mechanization, chemical fertilizers, and rapid distribution of fresh and processed food, the abundance of food in the United States is a success story. However, there are still Americans who are hungry, and there are starving people in many countries.

Activity 6-1
Always on a Diet

PART A: JOE'S DIET

"I'm always on a diet and I can't lose a pound," complained 200 lb Joe. "What foods are on your diet?"asks Betty. "Just the regular foods, but I skip lunch," he replies as he pops a Milk Dud into his mouth. Betty suggests that Joe keep a careful record of everything he eats for a whole day. Joe's record is on the chart on page 64. Using a calorie chart, calculate the total calories Joe consumed that day.

PART B: YOUR CALORIE INTAKE

Use the chart on page 65 to keep a record of the food you eat during a three-day weekend (Friday, Saturday and Sunday). This will include a school day and the free time on Saturday and Sunday. Add the calories for each food group in each time period to determine your total calories for each day.

After you watch the "Feeding the World" program, compare your caloric intake with that of the various groups mentioned in the program. What is the average intake of calories in your class?

Friday _____   Saturday _____   Sunday _____

Calories are one important thing to consider in the food you choose to eat, nutrients are also vital. From nutrition texts available in health courses or in the library, analyze the nutrients that you consumed in the three-day survey of your food intake. Compare these with the recommended nutrient diets found in these same books.

1. Which of the recommended nutrients did you neglect (leave out) in your diet of the three days studied? _____

2. How could you change your diet to improve the balance of nutrition in your food intake? _____

_____

_____

_____

3. If you were limited to foods from only three of the columns on your chart, which three columns would you choose? Why?

_____

_____

_____

JOE'S FOOD RECORD

Day	Food						
Friday Oct. 10	**Cereals, Grains**	**Meat, Fish, Poultry**	**Green Veget.**	**Potatoes**	**Dairy Products**	**Fruit**	**Sugars**
Waking 10 am	2 slices bread	2 slices bacon			2 eggs 8 oz milk	orange juice	
10 am– 2 pm	4 cookies				ice cream cone		
2– 6 pm	roll	¼ lb burger		french fries	milk shake		pie
6– Bed	4 cookies					banana	pie
Total Calories							

MY FOOD RECORD

	Cereals, Grains	Meat, Fish, Poultry	Green Vegetables	Potatoes	Dairy Products	Fruit	Sugars
FRIDAY							
6–10 am							
10 am–2 pm							
2–6 pm							
6–Bed							
Total Calories							
SATURDAY							
6–10 am							
10 am–2 pm							
2–6 pm							
6–Bed							
Total Calories							
SUNDAY							
6–10 am							
10 am–2 pm							
2–6 pm							
6–Bed							
Total Calories							

Activity 6-2
What's the Food Situation in Europe and Africa?

Countries with advancing technology of all types are increasing the amount of food produced per person. However, countries lacking in technological development are also those with the poorest agricultural development and the fastest population growth. The following data show the seriousness of this problem.

GRAIN PRODUCED PER PERSON (KILOGRAMS)

Year	Western Europe	Africa	Difference
1950	240	180	60
1955	260	180	80
1960	280	180	100
1965	300	175	125
1970	330	175	165
1975	400	175	225
1980	420	160	280
1985	500	150	350
1990			
1995			
2000			

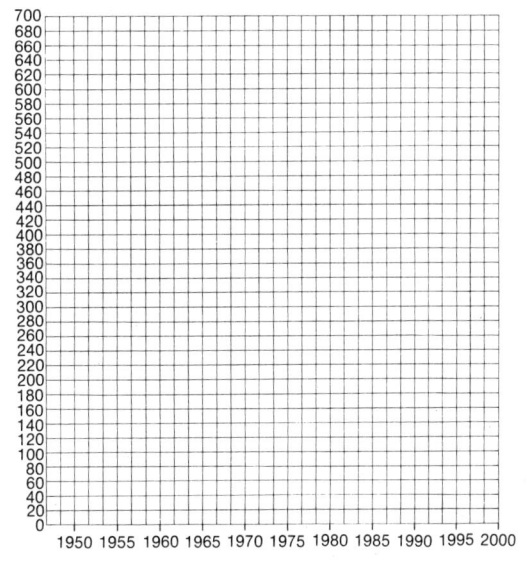

Fig. 6-1   Comparison of Grain Production Per Capita in Western Europe and Africa

PROCEDURE

1. Using Figure 6–1, plot the graphs for these three columns against the respective year. Use a different color for each column or area. Note that in 1950 there was not much difference between Western Europe and Africa. From your graph, predict the grain production per capita for Western Europe and Africa for 1990, 1995, and 2000. To do this, use dotted lines to extend the lines on your graph. Then read the expected new grain production from the points where the dotted lines cross the years 1990, 1995, and 2000. Put these numbers in your table below and indicate that they are estimates. Between the years 1950 and 1960 the population of Africa grew at the rate of 2% per year. Was there any increase in grain production? _____ If so, what was it? _____ How did you arrive at your answer? _____

   For the total period 1950–1985 the population of Africa doubled. By how much did grain production grow? _____ How did you arrive at your answer? _____

2. In groups of four students, answer the following questions: What technological systems could governments use in order to reverse the trend in per capita grain production? _____ Describe three such technologies. Discuss the trade-offs involved in putting each of them to use, and complete the following chart.

Technology	Trade-off
Increased Use of Fertilizer	**Possible Water Pollution**
1. _____	_____
2. _____	_____
3. _____	_____

Activity 6-3
Distributing Food

The availability of food for individuals varies greatly from region to region in this world. To demonstrate these differences in availability, you are asked to participate in this activity.

Your teacher will give you a colored card that represents a specific geographic area of the world. In the front of the room there are a number of items of food. These represent the total world food supply. When your color is called, you may go to that world food supply and take the amount which is now available in your area of the world.

When all areas of the world have received their quotas of food, all members of the class representing the same area (that is, have the same color of card) will meet in their respective groups to discuss and answer the following questions:

1. Which area of the world do you represent?
2. What is the population of that area?
3. How many students represent that area?
4. How many food items did each student get?
5. How do you feel about the amount of food the students received in relation to the population of that area? _____

    _____

    _____

6. How do you feel about the amount of food other groups received? _____

    _____

    _____

7. If the food had been distributed equally according to the population, how much would you have received? _____

    _____

8. What suggestions do you have as a representative of your area for more even distribution of food? _____

_____

_____

_____

_____

9. What are some of the reasons why some areas of the world have more food than others? _____

_____

_____

_____

_____

Activity 6-4
Saving Water While Growing Food (Drip Irrigation)

One of the problems of the countries in which there is widespread famine is that their water supply is not great enough to use the kind of spray irrigation commonly used in the United States. As a result, their crop yields are low. Because of continuing use without enough water, the soil dries up and blows away, leaving a desert. One answer to this problem is drip irrigation.

In this activity you will measure the amount of water used in growing a crop of radishes using drip irrigation as compared to growing the same sized crop using a sprinkling can as your irrigation system.

MATERIALS

- Two four-foot long window boxes (You can make these of wood or purchase plastic ones.)
- Packet of radish seeds
- Length of rubber tubing—the type used in science labs or purchased from auto supply stores
- Water "reservoir" with a spigot on the bottom (You can borrow it from the science lab or buy it from a store which sells them for home use in the refrigerator.)
- Sprinkling can
- Graduate cylinder for measuring the amount of water used
- Support for the "reservoir"

PROCEDURE

1. Read the directions on the seed packet to find how much space there should be between plants.

2. Drill small holes in the rubber tubing so that it will drip at the plant, but not between the plants. You will need to try a number of different size holes to get the right size. Start with a $^1/_{16}$" drill.

3. Allow enough undrilled hose to reach to the reservoir. Plug or clamp the other end of the hose at the end of the window box.

4. Fill the window boxes with potting soil.

5. Plant the seeds the same distance apart in each window box.

6. In one box, water the soil with the drip system until the area around the seeds is completely wet. Measure the amount of water used.

7. In the second box, water the soil with the sprinkler system until the entire surface of the soil is completely wet. Measure the amount of water used.

8. Put both boxes near a window so that they get the same amount of heat and light.

9. When the soil around the plants seems dry, water it with the correct system and measure the amount used.

10. When the crop is ready for picking, compare the size of the radishes from each farm. Cut the leaves off ½" from the top of the radishes and weigh them. Record all of your data in the following table.

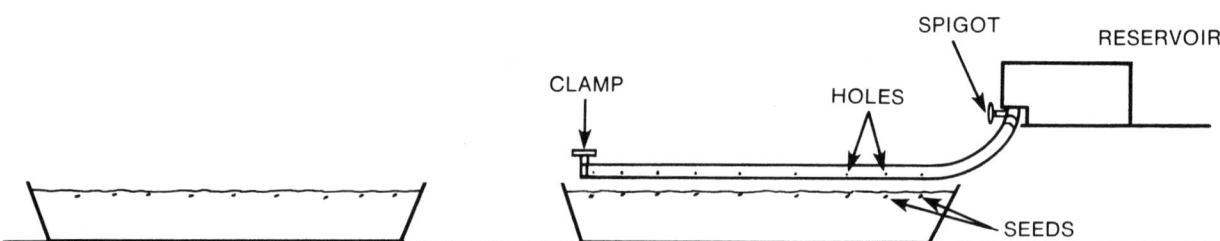

Fig. 6-2 Radish Farms

AMOUNT OF WATER USED

Date	Seeds		Plants		Weight of Crop	
	Drip	Spray	Drip	Spray	Drip	Spray

What was the difference in the total amount of water each in the drip and spray methods? _____

Which seeds sprouted first? _____

What was the difference in the weight of the crops? _____

How was this experiment different from the real-world situation?

_____

_____

PROGRAM 7

Communications: The Expanding World

The communications needs of human beings range from individual support, to community needs, to the business needs of international corporations. The telephone can be considered a communication system vital to all people in the United States today. The communications satellite offers instant on-the-spot news throughout the world. The possible effects of this instant communication on our behavior and values are hard to predict.

Name	Teacher
Class	Date

Activity 7-1
How Does the Fiber
in Fiber Optics Work?

When light passes from one medium (water) into another medium (air) it is bent (refracted). You can observe this by putting an object such as a coin in the bottom of a beaker or large glass. Look straight down at the coin. Now back away from the container so that you can no longer see the coin. Have your partner slowly pour water into the container until the container is almost full. What do you see?

Look at Figure 7–1. If a straight line is drawn from the edge of the coin on the left side of the figure, it misses the eye. You cannot see the coin. Draw straight lines in the figure on the right to show what you saw. Did the coin *seem* to have moved to the far side of the container? Since the coin did not move, and since light travels in straight lines, the light ray coming from the coin must have bent as it went from the water to the air.

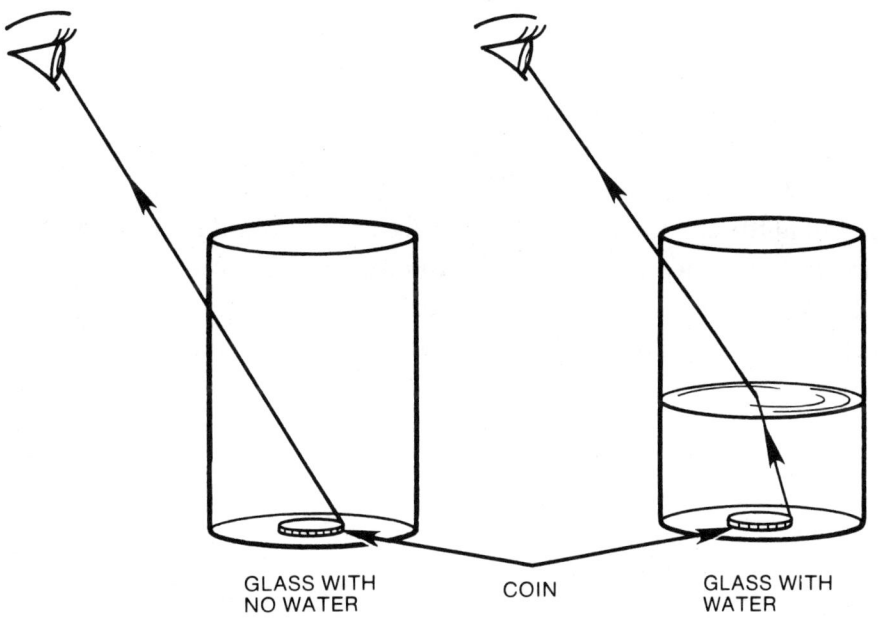

GLASS WITH NO WATER   COIN   GLASS WITH WATER

Fig. 7-1 Refraction of Light

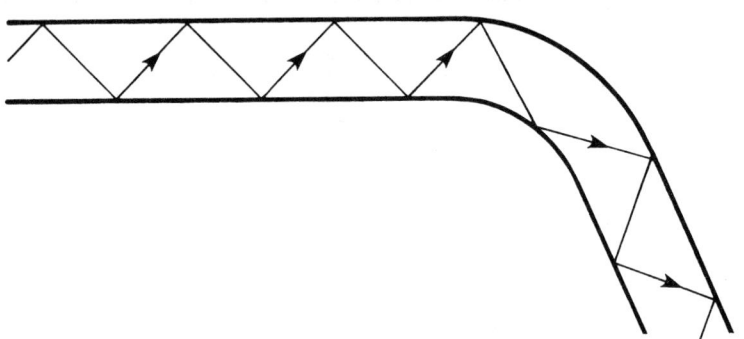

Fig. 7-2 Internal Reflection (The angle beyond which it is totally reflected back into the glass.)

Eyeglass lenses also bend light as it passes from the air into the glass and again as it passes from the glass back into the air. The amount of bending depends on the curvature of the glass and the angle at which the light strikes the glass. If the angle is very small, the light is reflected from the glass surface and does not enter the glass at all. This is called *total reflection*. The angle at which this takes place is the *critical angle*.

Try to determine this angle with a lens or even a flat glass plate. Remember that total reflection can take place as light passes from air-to-glass-to-air or from glass-to-air-to-glass.

Shine a flashlight on the polished end of a glass rod. The light will travel through the rod and come out the other end. Fiber optics work on the basis of total internal reflection. A beam of light enters one end of a very thin fiber of glass (about the thickness of a hair) and cannot leak out the sides of the fiber. The light hits at an angle beyond the critical angle and is reflected back into the glass fiber as shown in Figure 7–2.

Note: Different materials have different critical angles. For water this value is 49 degrees from the perpendicular to the surface; for glass, 41.3 degrees, for diamond, 24.4 degrees. This means that any light ray which hits the inner surface of a face of a diamond at an angle greater than 24.4 degrees from the perpendicular will be internally reflected and produce the sparkle of diamonds. When cut at the correct angle a diamond can focus light into a very strong beam. This is what the enemy in the James Bond movie *Diamonds are Forever* was planning to do.

Since glass has a critical angle of 41.3 degrees, it is only necessary to shine a light into the polished end of the glass and to have no bends in the glass tube, which would provide an angle less than 41.3 degrees until we get the light to where we want it to go. Light travels as an electromagnetic wave, just as radio waves do. Therefore, a message can be carried by the light wave just as a message can be carried by a radio wave.

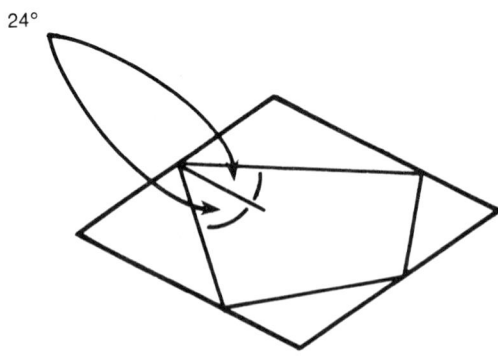

Fig. 7-3 Internal Reflection of Light in a Diamond

Activity 7-2
Sending Messages
Through Space

Just as dots and dashes can stand for letters in Morse code (·_ = A, _··· = B, _·_· = C, _·· = D, · = E), we can make up a code of ones and zeros to stand for letters. A message can now be sent into space as ones and zeros (a digital code), received by a satellite, and retransmitted to a spot on earth that could not be reached by radio waves traveling a straight line, Figure 7–4.

You can make a digital coder. Using your digital coder you will send the message by wire to a light. The light can stand for a microwave transmitter or laser light source.

Cut a strip of oak tag about 1″ wide. Punch holes to stand for a three-bit code. A hole stands for a "1" in the code. A blank space at the spot where a hole could be stands for a "0". The three-bit code represented in Figure 7–5 is 110 111 001 011.

If 000 stands for A and 001 stands for B, complete the rest of the alphabet that could be represented by your three-bit code. What size bit code would be required to complete the alphabet?

000=A   001=B   _____ = C   _____ = D   _____ = E   _____ = F   _____

_____

_____

_____

The message in the sample is _____ .

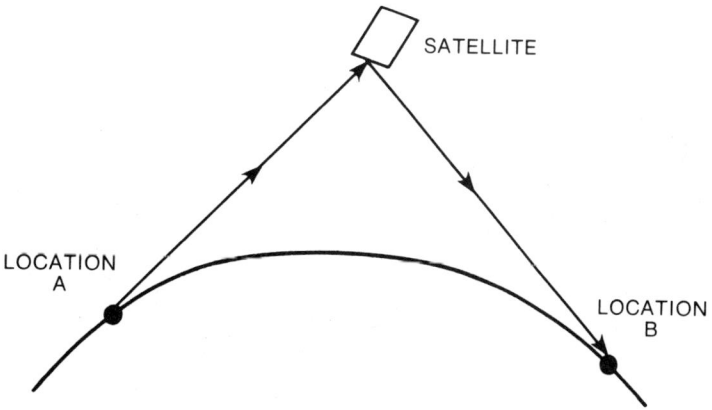

Fig. 7-4 Radio Waves Travel in Straight Lines

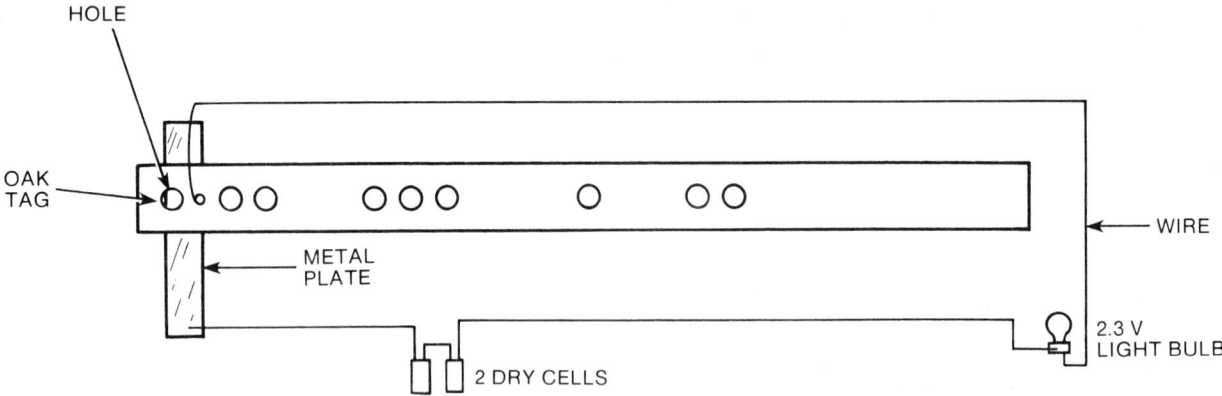

Fig. 7-5   Digital Coder

This will be different for different people unless they use the same code for each letter. Agree with your partner (the receiver on your code) on which three-bit group stands for which letter. Each bit requires the same space. The sample uses three blocks/bit. As you pull the strip past the metal plate at a constant speed (one space/second), the wire will touch the plate when a hole appears under the wire. The circuit will be completed and the light will light. The receiver will record this as a "1." If in the next time interval no light is seen, the receiver will record a zero, and so on. When you first start to send your message, it might be a good idea to pause after each three bits are sent. When you have finished sending your message, compare what you sent with what your partner thinks was received. How well did you do?

Change positions with your partner as sender and receiver. Use more strips and punch new codes. Remember to space the bits evenly on your strip.

Paychecks, magazine subscriptions, and computer programs were all recorded on punched cards at one time. With the improvement of Optical Scan Systems, the punched card and punched tape were no longer used. However, the binary code that was used on the punched card to give the message is still used. The Bar Code |||||||||| on consumer items is also based

0 ‖288762‖ 3

on this same binary code. The width of the bar or blank space stands for the number of "1"s and "0"s.

The beautiful pictures of other planets sent back from outer space are sent back as groups of "1"s and "0"s. These are decoded by computers on earth and reproduced as pictures. Digital discs for hi-fi and TV are also coded sets of "1"s and "0"s to stand for both sound and picture.

Activity 7-3
Communication from Human to Human

The development of the technological systems for instant communication from any spot on the earth to any other spot has many advantages. For example, we can see sports events or new events from anywhere on earth just as they happen. Government officials from one country can talk to officials from another country instantaneously. As with many technological developments, there are also problems. One of those problems is that humans do not always say exactly what they mean. In many areas of life, that lack of precision can cause problems. This activity will give you a chance to test your ability to say what you mean.

Your teacher will give your group some materials with which to build some device. After you have built it, have one person in the group draw a diagram of the device while the others work together to write a detailed description of how to build it.

When you have finished drawing and writing, take the device apart. Present the pieces and written description to the teacher. Keep the diagram. Do not show it to any other group.

The teacher will give your pieces and description to another group so that they can follow your directions and build the same device. You will receive the directions and parts to build a device planned by someone else.

When all of the devices have been built according to the written directions, they will be returned to their original designers.

You will compare your diagram of the original device with the device built by the other group. How close are they to each other? Did you leave out some instructions? Did the other group misunderstand your directions? Were you able to build the device from the directions given by another group?

Bring in a set of instructions on how to put together something that was bought from a store. Find any errors. Rewrite the instructions on the basis of what you have learned in this activity.

What problems can be see resulting from a misunderstanding of human-to-human communication?

1. On a person-to-person level _____

_____

_____

2. On a person-to-government level _____

_____

_____

3. On a government-to-government level _____

_____

_____

PROGRAM 8

The Changing Romance: Americans and Wheels

The fondness of Americans for automobiles shows their love of mobility and freedom. Automobiles were first owned only by wealthy people until Henry Ford had the idea that working people should also be able to afford one. Ford developed the method of assembly line production which made automobiles at such a low cost that almost every working American could buy one. The use of the assembly line spread through the manufacturing industry and made the United States the most productive and wealthiest nation in the world.

The automobile has costs as well as benefits. It has produced urban sprawl (unplanned expansion of a city) and the suburb, while polluting the air and land. Using private automobiles rather than public transportation is proving very costly both in money and environmental damage. We are always being challenged to improve the trade-offs.

Activity 8-1
Traffic Delays

Traffic delays are often caused by the design of roads and traffic flow patterns for the state, county, and community. These designs may have been good at one time, but are not good now. This situation may be true in the halls of your school. A good traffic flow pattern for Tuesday at 8 am may not work for Friday at 2 pm.

The purposes of this experiment is to look at the traffic flow in your school for a specific time. Then you can make predictions for other times and possibly reduce problems involved with the traffic flow. You and your classmates will measure traffic density (number of people passing a given spot in a given direction each minute) during the change of class just before the class in which you are doing this analysis. From the data collected and the model developed, predictions and possible solutions to passing problems may be made.

PROCEDURE

1. Get a floor plan of the school.

2. From your assigned station at the intersection, count the number of students passing that intersection in the given direction each minute, during the time just before your class meeting. Your teacher will need to get permission for you to leave one class a little early. If you are stationed near an intersection close to your other classroom, this delay can be kept to a minimum.

3. Figure 8–1 shows the traffic monitor assignments. Students at positions 1, 3, 5, and 7 will count the number of students who enter the intersection each minute from their hallway. Students at positions 2, 4, 6, and 8 will count the number of students who leave the intersection each minute.

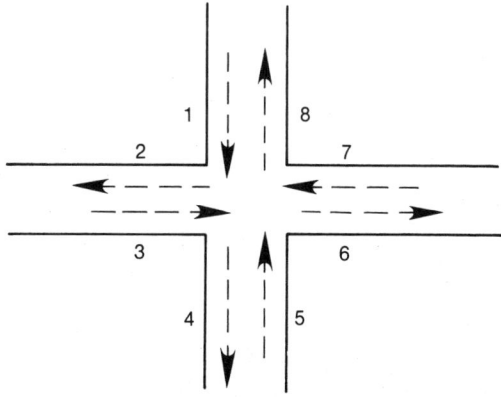

Fig. 8-1   Traffic Monitor Assignments

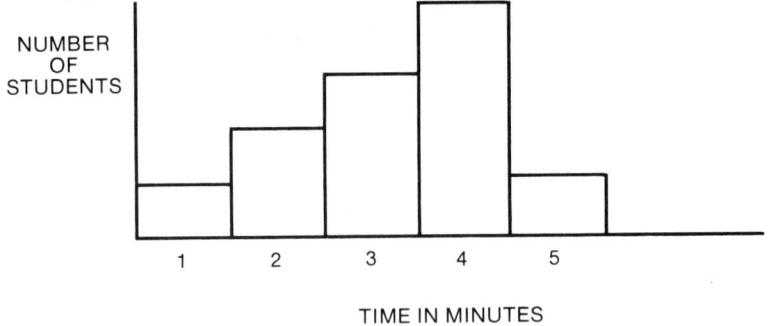

TIME IN MINUTES

Fig. 8-2   Traffic Density at One Traffic Position

4. Each person will construct a bar graph of the number of people counted for each passing period. The graph will probably look something like Figure 8–2.

5. On a single floor plan of the shcool, display the traffic flow for that specific passing time. On a piece of paper, record the number of people passing in each direction for each of the minutes in the period between classes. Figure 8–3 is an example of a possible traffic pattern. The number in parentheses is the minute that the highest number of people were moving through the intersection. The number 70 (4) shows that 70 was the highest number of people walking east at the fourth minute of the period. Record your data in the space provided.

   An analysis of the numbers in Figure 8–3 shows that a total of 180 students entered and left the intersection during the five-minute passing period, and that 70 of them left in the fourth minute in the same direction. Maybe that is why so many students are late for classes along that hallway before Period 3.

6. With this information and the master schedule of the school, predict the traffic flow patterns and possible problems during other passing periods.

7. If your teacher and principal know of problems at different times during the day, study the traffic flow during that time.

8. Your group or your whole class may find other solutions. While overpasses and entrance and exit ramps are not part of the solution to this problem, some other alternatives might be.

This type of study has been made in other schools and has improved traffic flow in the school. Studies like these can be made of the flow of street traffic around the school just before school opens and closes.

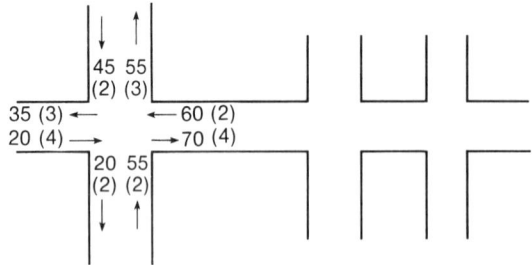

Fig. 8-3   Sample: School Traffic Between Periods 2 and 3 (Maximum number of people a minute is in parentheses.)

Activity 8-2
Traffic Problems in the City

It has often been said that problems that face New York City today will face every large city in the future. Listed on pages 88 and 89 are two of New York City's auto-related problems with suggested solutions.

Your group of four students will be assigned one of the problems. For your problem, you are to:

- Study the problem and the suggested solutions.
- Consider and record added solutions from your group.
- Discuss and record problems that might result from each solution.
- Agree on the trade-off that might be the best way to solve the problem. Record your choice.

Report your decisions to the whole class with the other groups' reports. Record the other groups' solutions, problems, and trade-offs.

Problem 1. Severe traffic congestion entering the city between 8 am and 10 am and leaving between 4 pm and 6 pm.

Suggested Solutions:

1. Have businesses stagger their starting and quitting times.
2. Charge tolls on all bridges and tunnels entering the city from the suburbs.
3. Set tolls at $1.00 for cars with three people, $2.00 for cars with two people, and $3.00 for cars with only the driver.

PROBLEM 1—WORKSHEET

Additional Solutions	Resulting Problems	Trade-offs	Choice

Problem 2. Heavy pollution from auto exhaust especially on days of no wind and bright sunshine.

Suggested Solutions:

1. Encourage bicycle driving in the city by marking bicycle lanes on all streets.
2. Allow no street parking between 7 am and 7 pm.
3. Increase parking garage rates to $5/hour between 7 am and 7 pm.
4. Allow only electric-powered cars to use the city streets between 7 am and 7 pm.

PROBLEM 2—WORKSHEET

Additional Solutions	Resulting Problems	Trade-offs	Choice

After completing this activity, check the papers and the police in your area to learn about problems related to auto traffic. Use the space below to write any suggested solutions. Are they workable?

Activity 8-3
Auto Safety

In the section of the program dealing with auto safety the following items were mentioned:

- Air bags
- Mandatory seat belts
- 55 mph speed limit
- Drunk driving

- Safe speed
- Insurance
- Cost of repairs

Late in the video program, a statement is made that all the systems within the transportation system are related to each other. In this activity you will measure one part of the human-machine interaction. You will then discuss how this part is related to each of the items at the beginning of this activity.

HUMAN REACTION TIME

There are many different ways of measuring human reaction time. A simple computer program can be made to do this. A system of a light, two switches, and a timer can also be arranged. The circuit shown in Figure 8–4 uses a switch operated by a person who is behind the person being tested.

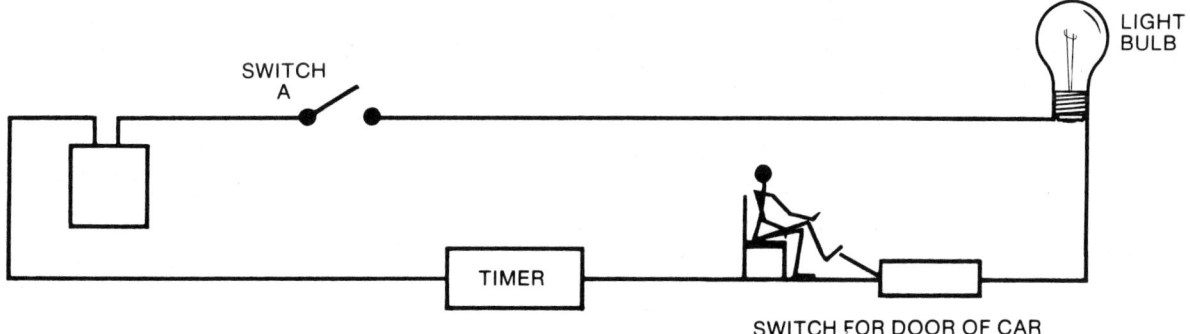

Fig. 8-4   Reaction Timer

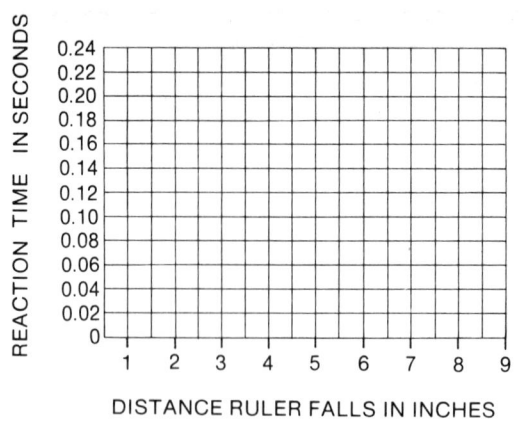

Fig. 8-5   Reaction Time Found from the Distance the Ruler Falls

The person being tested has his or her feet on the floor. When switch A is closed, the light that stands for a traffic light changing from green to yellow goes on. At the same time, the timer starts. When the "driver" sees the light go on, he or she lifts the right foot from the floor and pushes on the brake, turning off the light and stopping the timer. The elapsed time (the time it takes) is the driver's reaction time.

A simpler but less realistic way to measure reaction time is to have one person hold a ruler vertically just above the fingers of the "driver." When the ruler is dropped, the "driver" catches it between thumb and finger. Note the distance the ruler fell by the point at which it is caught. Using the graph in Figure 8–5, read across the horizontal axis to find the distance the ruler fell. Where this line crosses the graph read to the left to find the time elapsed. This is the driver's reaction time.

Which system of measuring reaction time is closer to the real situation of driving a car?

_____

Which system would you expect to show a longer reaction time? _____

_____

The actual reaction time is obtained from repeated testing of drivers in lifelike simulations in which the time that the right foot takes to move from a gas pedal to a brake pedal varies. The range is from 0.8 second for an average well-coordinated person to 1.0 second for the person who is not paying strict attention to the lights to 1.6 seconds for the person with a blood alcohol content (BAC) of 0.06 percent (not legally drunk). The legally drunk BAC in most states is 0.10 percent. Reaction time for this situation varies greatly from person to person.

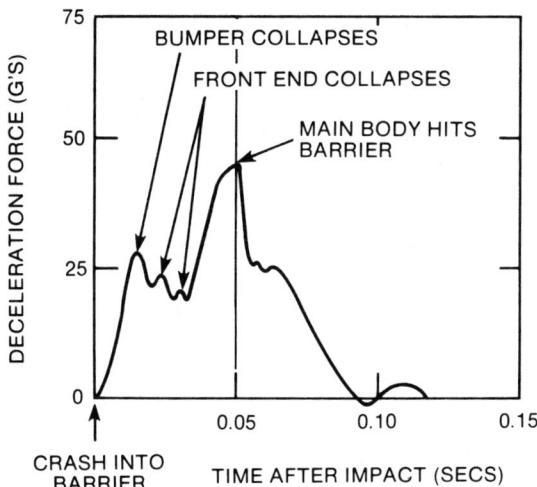

Fig. 8-6 Deceleration Force on a (1000 kg or 2000 lb) Car Whose Impact Speed Is 66 km/h (40 mph)

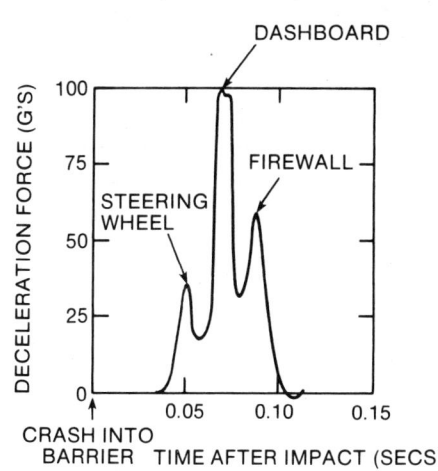

Fig. 8-7 Deceleration Force on an Unrestrained Dummy Whose Mass Is 76 kg (165 lb) When the Impact Speed is 66 km/h (40 mph)

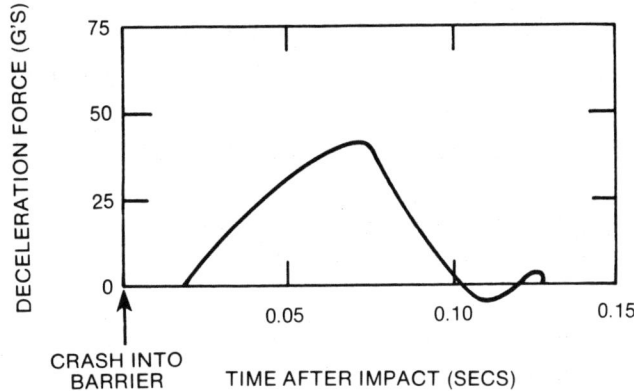

Fig. 8-8 Deceleration Force on an Unrestrained Dummy Whose Mass Is 75 kg (165 lb) When the Impact Speed Is 66 km/h (40 mph)

Figures 8–6, 8–7, and 8–8 were plotted from data derived from controlled crashes of actual automobiles in which human-shaped dummies were placed both with and without seat belts.

PROCEDURE

1. Look at the graphs in Figures 8–6, 8–7, and 8–8 with respect to your reaction time. Could you brace yourself fast enough in a 40 mph crash? In Figure 8–6, the car traveling at 40 mph begins to slow down about 0.01 second after it hits the barrier. Since your reaction time is about 0.8 second you have not yet started to raise your arms. Your head is still moving forward at 40 mph or 27 ft/second; it is 0.27 foot or 3.25 inches closer to the windshield. At 0.05 second, you still have not raised your arms and your head has now moved _____ inches closer to the windshield. At 0.8 second your head has moved _____ inches forward.

   In Figure 8–7, note the points where your head hits the steering wheel and where your passenger's head hits the dashboard and firewall (the firewall is directly in line with the windshield). A force of 100 g's is 100 times the weight of the object. If your head weighs 15 pounds, it hits the dashboard with a force of _____ lb, and the firewall (windshield) with a force of _____ lb.

2. The air bag opens completely in 0.02 second. How far has your head traveled in that time? _____ feet   _____ inches

3. Shoulder harnesses protect us in side crashes, roll-overs, and head-on crashes. What laws should be passed related to air bags and shoulder harnesses?

   _____

4. What problems might arise if air bags are finally installed in all cars?

   _____

   _____

5. How can these problems be overcome? _____

   _____

   _____

   _____

6. At 55 mph your car is traveling 73 ft/second. How far should you stay behind the car in front for minimum safety while traveling at 55 mph (Figure 8–5)? If your reaction time is 1 second then: 1 sec × 73 ft/second = _____

7. Engineers often build a factor of safety into their buildings, bleachers, and bridges. The factor of safety is the amount you multiply the minimum safe system by to assure safety. If you wanted a factor of safety of 2 for driving behind a car at 55 mph how far away should you stay? _____
   It is hard to estimate distances while driving, especially since speeds change. If you assume average reaction time of one second, then two seconds allows a

realistic safe distance. The way you figure this out is to watch when the car in front of you passes a landmark (pole, mark in road, signpost), start counting one thousand one, one thousand two. If you get to that same spot before you say two, you are too close. Why does this system work for all speeds?

_____

_____

8. If your car travels 73 feet in one second at 55 mph, how many feet will your car travel in one second at 70 mph? _____

_____

How far will it travel during a reaction time of 1.2 seconds at 70 mph?

_____

PROGRAM 9

China, Japan, and the West

When Westerners first visited China in the 1400's, they found that Chinese technology was well developed. The technology transfer was from China to the West. But the levels of technologies were reversed when Americans visited China in the 1800's. The development of technologies is strongly affected by the philosophy and expectations of the cultures. When the Japanese changed their expectations of technology in the 1900's, their response changed completely from their early history.

Activity 9-1
Measuring Time

The technology which is seen at the beginning and the end of the program "China, Japan, and the West" is the stopwatch. Stopwatches can measure very small differences in time. Olympic races are won and lost in a thousandth of a second. The human ability to tell time is rather limited, and yet with practice can become quite accurate for short amounts of time. In this activity you will have the chance to find out: (a) how accurately you can guess a minute of time; and (b) how accurately you can measure time in fractions of a second.

At a given signal, your teacher will tell you to begin estimating a minute of time. You will do this with your eyes closed and without counting. When your teacher says "Go" you will begin estimating how long a minute is. When you think a minute has gone by, you will raise your hand and open your eyes. If there is a clock with a second hand available, note the position of the second hand and record the time.

Example:

TRIAL	ACTUAL TIME	MY TIME	MY % ERROR
1	60 sec	40 sec	33%
2	90 sec	100 sec	11%
3	130 sec	110 sec	_____
		average % error	_____

For Trial #1 your error is: 20 seconds − 20/60 = 0.33 or 33%.
For Trial #2 your error is: 10 seconds + 10/90 = 0.11 or 11%.

Calculate the error + % error for trial 3. Add the percent errors and divide by 3 to get the average percent error.

TRIAL	ACTUAL TIME	MY ESTIMATE	MY % ERROR
1	_____	_____	_____
2	_____	_____	_____
3	_____	_____	_____
		average % error	_____

After three trials for different lengths of time you will calculate your percentage of error and tell whether it was plus or minus.

You can use these calculation techniques for the following experiment.

In this next activity, you will try to measure fractions of a second. When the teacher says "Go," start tapping your pencil on a clean sheet of paper to make dots. At the end of ten seconds, the teacher will say, "Stop." Count the number of dots you have made. Divide this number by ten. This tells the number of dots per second. Record this information in the table.

TRIAL	TIME	DOTS	DOTS/SEC	TIME BETWEEN DOTS
Example	10 sec	56	5.6	$1/56$ = 0.17 sec
1	10 sec	_____	_____	_____
2	10 sec	_____	_____	_____
3	10 sec	_____	_____	_____
4	10 sec	_____	_____	_____
5	10 sec	_____	_____	_____

During the next four trials, try to control the speed at which you make the dots so that you make 50 dots in 10 seconds. You now have a built-in stopwatch that measures about 0.2 second. Use it to time the following activities.

In your group of five, have each person throw a wad of paper straight up. Start timing when the paper leaves the thrower's hand. Stop timing when the paper hits the floor. Count the dots and divide by 5. This will be your time for the event. Compare this with the average time of the other four.

SAMPLE		ACTUAL TIME	
My time	$7/5$ = 1.4 sec	My Time	_____
Partner 1	1.2	Partner 1	_____
2	1.6	2	_____
3	1.0	3	_____
4	1.6	4	_____
Total	6 8 sec	Total	_____
Ave = Total/5 = 1.36		Average	_____

What other events could be timed this way? _____

_____

_____

_____

If time is available try them.

Activity 9-2
Technology Assessment

Assume the following situation: The stern rudder, the compass, printing with movable type, and gunpowder have been invented by the emperor's chief inventors. The people of China know nothing about these inventions.

The emperor has called in his chief advisors (as many as you have in your class). He has divided this group of advisors into four subgroups. Each subgroup has been told about one of the inventions and has been commanded to make a technological assessment of the possible effects of these inventions. Your group's recommendations are due in thirty minutes. Your group has been assigned the technology assessment on _____ .
What are some of the possible results of that technology?

1. _____

2. _____

3. _____

4. _____

How probable is it that the suggested result would occur?

	Very Probable	Not Probable	Impossible
1.	_____	_____	_____
2.	_____	_____	_____
3.	_____	_____	_____
4.	_____	_____	_____

What effect would that result have on Chinese civilization? Effect of each result:

1. _____

2. _____

3. _____

4. _____

What did the groups recommend to the emperor regarding each technology?

Stern rudder _____

_____

Compass _____

_____

Printing with movable type _____

_____

Gunpowder _____

_____

Name		Teacher	
Class		Date	

Activity 9-3
Japanese Quality Techniques

One of the reasons given for the ability of the Japanese to make quality products at less cost than the cost for identical products made in the United States is that they organize their work better.

In this activity, you and your team members will be given a job to do. It might be sorting cards, nuts and bolts, or books, or any other set of items. It might be putting together a structure from a set of parts such as a Tinker Toy® set or Structo®. Your team's ability to do this most efficiently will be based on the time it takes to do the job and the accuracy with which you do it.

SORTING TASK

When you are given the items to be sorted, <u>but before touching any of them,</u> meet with your group. Be sure that everyone knows what is to be done and who is responsible for each task.

Name

1. _____
2. _____
3. _____
4. _____
5. _____

Responsibility

1. _____
2. _____
3. _____
4. _____
5. _____

RESULTS OF SORTING TASK

Trial	Time	Errors
1	_____	_____
2	_____	_____
3	_____	_____
4	_____	_____
5	_____	_____
Average	_____	_____

RESULTS OF SORTING TASK

Group	Average Time	Average Errors
1	_____	_____
2	_____	_____
3	_____	_____
4	_____	_____
5	_____	_____
6	_____	_____

CONSTRUCTION TASK

In this task your team of six people will join with another team of six people in a competition to design and build a structure from a set of parts that will be given out by the teacher. The other teams of six will be given the same assortment of parts.

In each team, three people will be assigned the task of designing and writing instructions for assembly of the structure. The other three people *use the design and written instructions only* to build the structure. The two groups may not talk with each other. Assume that each group is in a different building, the design building and the assembly building. When the assembly group has completed its structure it will be compared with the original design.

Team competition will be judged on these items on a scale of 0–10, 10 being the best.

Time used in design
Time used in construction
Usefulness of structure
Accuracy of construction

	TEAM SCORES			
Team	Design Time	Construction	Usefulness	Accuracy
1.	_____	_____	_____	_____
2.	_____	_____	_____	_____
3.	_____	_____	_____	_____
4.	_____	_____	_____	_____
5.	_____	_____	_____	_____
6.	_____	_____	_____	_____

PROGRAM 10

Population Patterns

Most people would agree that it is good to reduce the death rate, to extend life, and to have children. However, the startling increase in the rate of growth of the world population has produced pressures on our material and energy resources. These pressures are so great that attitudes and customs are being reevaluated because of the new alternatives that technology can provide.

Activity 10-1
Really Big Numbers

In the "Population Patterns" program there are numbers like five million, four million, etc. It is hard to imagine numbers greater than 10,000. For example, we know that it is possible to buy a car for $10,000. Some cars cost more than $10,000 and some cost less. When we use numbers of even 100,000, we have trouble relating that to the real world. In terms of cars that would be ten $10,000 cars or five $20,000 cars. In order to sell one line of luxury cars, one manufacturer tells on TV ads that for the $70,000 price tag on a foreign luxury car, you could buy the best of his cars and still have money left over to hire a chauffeur for a year.

In this activity, you will have the chance to calculate or collect one million of something. The sheet of graph paper (Figure 10–1) will be used as the basis for your calculation. The marked-off area of the paper contains 40 squares in one direction and 50 squares in the other direction.

How many square are there in the marked-off area? _____

How many sheets of similar graph paper would you need to have to make up 1,000,000 squares?

_____

1,000,000 divided by _____ squares/sheet = _____ sheets.

A ream (package) of paper for the copying machine has 500 sheets. How many reams would you need? _____

Each minute has 60 seconds and each hour has 60 minutes. How much time in days _____ hours _____ and minutes _____ = 1,000,000 seconds?

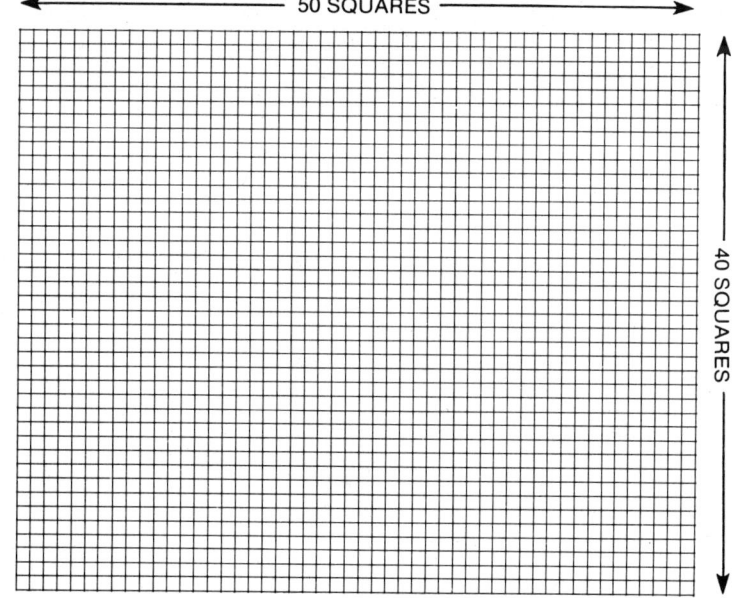

Fig. 10-1   Step One. Really Big Numbers

Suppose you won one million dollars in a lottery and the money was awarded to you at the rate of $100/day. How old would you be when you collected the final $100?

_____ years old

Have four students stand in pairs back to back to form a square. Measure the floor area taken up by the four students.

_____ ft × _____ ft = _____ square feet.

Let's assume 10 square feet for the four students. This would be 2.5 square feet per person. If 1,000,000 people this size crowded onto a highway 100 feet wide, for how many miles would this crowd stretch? Remember: one mile equals 5,280 feet.

Each of these examples were included to help you picture the size of 1,000,000. Think up two other examples that would help people visualize 1,000,000.

Now that you have an idea of how large one million is, let us look at one billion (1,000,000,000). Suppose in that lottery problem you had won one billion dollars and collected it at the rate of $1,000/day, or $365,000/year. How old would you be when you have collected all of the $1,000,000,000?

The world population in 1950 was 2,500,000,000 (2.5 billion). In 1985 it was 4,800,000,000 (4.8 billion). What was the average number of people added to the population each year for that 35 years?

_____ people/year

How many people live in your town or city? _____

How many more communities the size of yours have been added to the earth from 1950 to 1985?
_____

2,500,000,000 divided by the number of people in your community = _____ of communities.

Since we do not want to increase the death rate, the only way to slow down the growth of population is to _____

_____

Activity 10-2
Conflicts From Growing Numbers

One of the main causes of conflict between different cultures living in the same country is that differences in population growth rates of the groups can result in the minority group of people growing in numbers until they become the majority group.

For example: Two countries, A and B, border each other. For economic reasons, country A with a population of 50 million defeats country B, population 30 million, in a war. The two countries are now combined into one. The new country has a democratic form of government and all citizens As and Bs are free to vote. The growth rate of citizens from country A is 8% every 10 years. The growth rate of citizens from country B is 15% every 10 years. In how many years will the number of citizens from country B be equal to the number of citizens in country A? Completing Table 10–1 will help you answer this question.

TABLE 10-1   RESULTS OF DIFFERING GROWTH RATES OF TWO POPULATIONS

Year	Population A in Millions	Population B in Millions
1990	50	30
2000	50 × 1.08 = 54	30 × 1.15 = 34.5
2010	54 × 1.08 =	34.5 × 1.15 =
2020	× 1.08	× 1.15
2030	× 1.08	× 1.15
2040	× 1.08	× 1.15
2050	× 1.08	× 1.15
2060	× 1.08	× 1.15
2070	× 1.08	× 1.15
2080	× 1.08	× 1.15
2090	× 1.08	× 1.15
2100	× 1.08	× 1.15

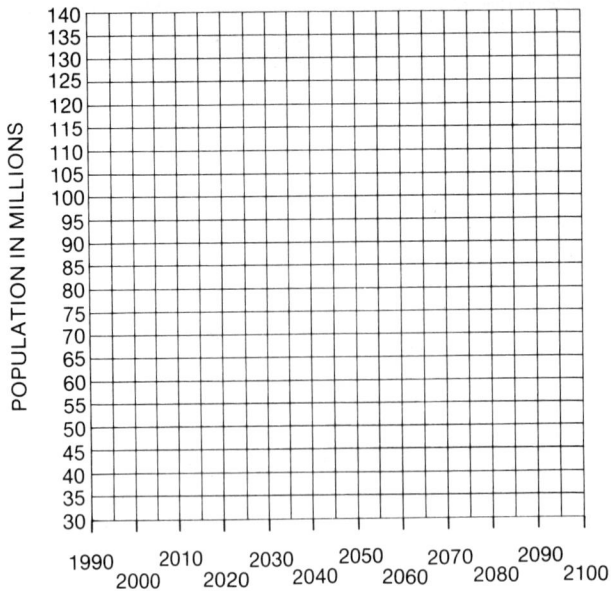

Fig. 10-2   Results of Differing Growth Rates of Two Populations

Use Table 10–1 and Figure 10–2 to figure out the growth of A and B citizen populations for the next 100 years. On the graph, show how these growth rates differ from each other. You will be assigned to calculate the population growth of one of the populations. Your partner will calculate the growth of the other population. The way to get the population for the year 2000 is to multiply the population in 1990 by 1.08 if the population is growing at the rate of 8% every ten years, or by 1.15 if the population is growing at the rate of 15% every ten years. For future years, multiply your previous census by 1.08 or 1.15, depending on which population you represent. A calculator is a great help here. Continue this calculation for each ten years up through the year 2100. Use Table 10–1 to record the calculations. Then plot that data on the graph in Figure 10–2.

To compare the growth of populations A and B, plot the results of your calculations on the graph in Figure 10–2. Then your partner can plot the results of the calculations for the other population.

As the democratically elected government of this new country begins to see the population of the minority group catching up to their own, they see that their majority is becoming the minority. Discuss some strategies that they might begin to use in the year 2010 to protect the majority status. These strategies must be humane and fair to both sides in order to pass the criteria of their constitution and value system.

Each strategy suggested must be measured against three criteria:

1. Is it humane?
2. Is it fair?
3. Will it work?

Are there any examples in the past or the present that are similar to the situation present in this activity? If so how were, or are, they handled?

Activity 10-3
Reducing Population Growth Rate

The program uses the example of a deer population growing so fast that almost all of the available food is used. Meet in a group of four students. You will be assigned one of the following topics to discuss in your group:

1. A world situation that is similar to the deer situation.
2. Technologies that humans have to reduce the danger suggested by the example of the deer in Arizona.
3. The effect of modern transportation on the population growth through transmission of disease, delivery of medicine, and any other effects.
4. The effect of improved education and industrialization on maintaining a livable population growth.

After you complete your discussion, your group and the other three groups will report their conclusion to the class. Take careful notes.
NOTES:

After studying the notes on each of the topics discussed, write your recommendations to the next International Conference on Population.

RECOMMENDATIONS:

PROGRAM 11

Exploring Space

Throughout history, humans have wondered about "What is out there?" With telescopes, we studied the solar system and the stars. Now we actively explore space with rockets. Development of rocket technology during World War II has led to the space race between the United States and Russia. Space shuttles and instrumented exploration of the planets has given us data and spectacular photographs of our solar system and about our galaxy. Now a major question in space exploration is to decide when to send human beings and when to send instruments or robots.

Activity 11-1
Orbit Size Compared to Speed

In the program, satellite orbits close (100–300 miles) to the earth (shuttle, weather satellites) are discussed along with geosynchronous orbits (24 hours) that are about 26,000 miles above the earth and the faster the satellite travels, the larger will be its orbit; the larger the orbit, the smaller the gravitational force between the earth and the satellite. If the satellite goes fast enough (25,000 mph), its orbit will be so large that the gravitational pull will be practically zero and the satellite will go off into space. Satellites that explore other planets or even go to the moon must first leave earth's gravity.

You can demonstrate the relation of speed to size of orbit using the following materials:

- Glass tubing from the chemistry lab, about 6" long
- String about 18 inches long
- Weights to tie on the string
- One-hole stopper as the satellite

Thread the string through the glass tubing. Tie a weight to one end of the string. Tie a one-hole rubber stopper to the other end of the string. With the tubing in a vertical position as in Figure 11-1, start to rotate it so that the rubber stopper tied to the end of the string which comes out of the top of the glass tubing begins to go around in a circle.

What happens to the size of the orbit as you make the satellite travel faster? _____
_____
slower? _____

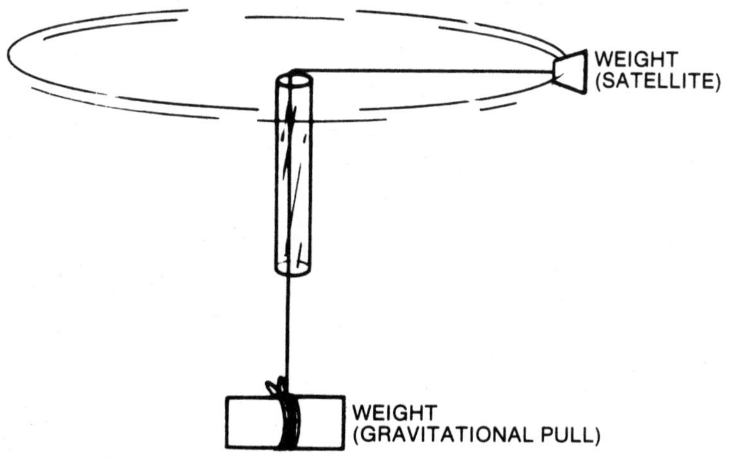

Fig. 11-1   Satellite Simulator

119

Tie an added weight on the bottom weight of the string. Repeat the experiment. Record your observations as you try to keep the same size orbit as in the previous trial.

_____

_____

_____

The bottom weight and the string combined represent the gravitational pull between the satellite and the planet around which it is orbiting.

If you did this experiment outside and the glass tube was large enough in diameter for the weight to pass through it, what could you do to have the "satellite" leave the planet's gravitational field? _____

_____

What did you observe in your classroom experiment to lead you to the answer to the preceding question? _____

_____

Assume that the planet (top of the glass tube) is rotating at the rate of one turn per second (10 revolutions in 10 seconds). If you want to keep the planet in geosynchronous orbit, how fast would it have to go? Try it. When you get it to the right speed, grasp the string just where it comes out of the bottom of the tube and allow the satellite to stop. Now measure the length of the string from the top of the tube to the planet.

To find the speed of the planet, first you need to calculate the circumference of the orbit, $C = 2 \pi r$, then divide the circumference by the time for one orbit.

Repeat for a different time of rotation; for example, 0.8 second (8 orbits in 10 seconds) or any other time of rotation. Record your data.

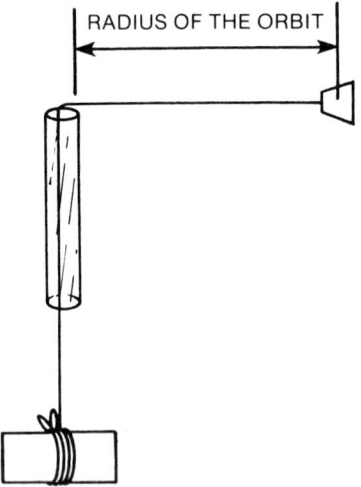

RADIUS OF THE ORBIT

Fig. 11-2   Radius of Geosynchronous Orbit

	TRIALS		
Trial	Time of Orbit	Radius of Orbit	Speed of Orbit
1			
2			
3			
4			
5			

On the graph in Figure 11–3, plot the radius of the orbit against the time for one complete revolution.

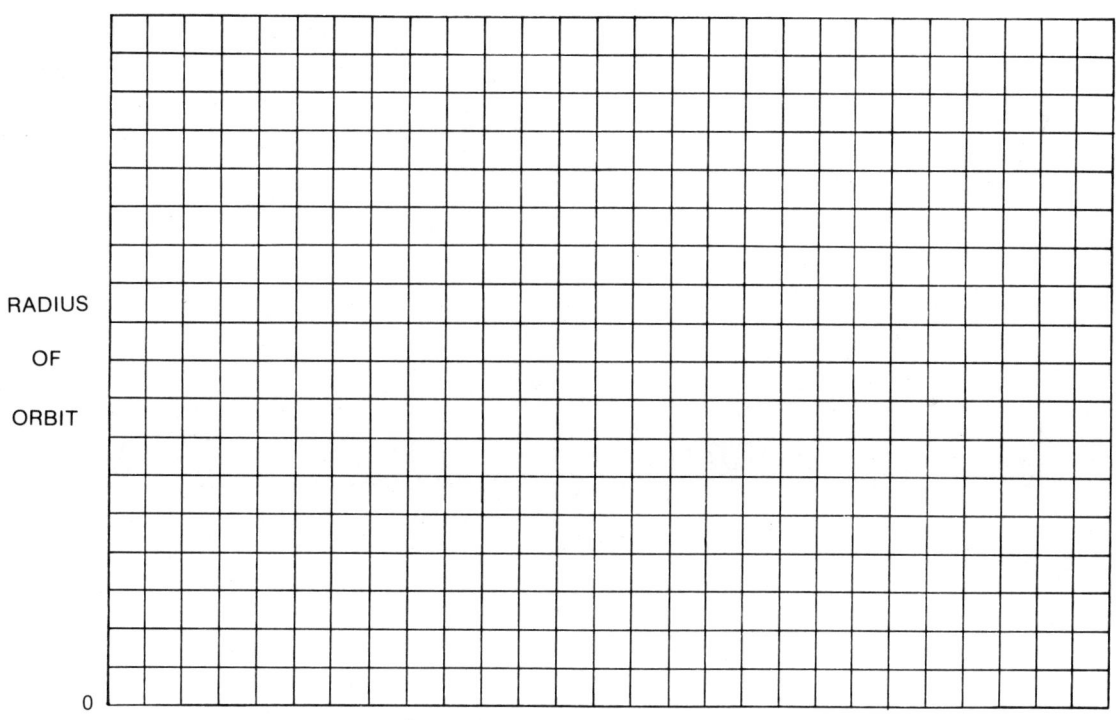

RADIUS

OF

ORBIT

0

GEOSYNCHRONOUS TIME IN SECONDS

Fig. 11-3   Radius of Orbit vs Time for One Complete Revolution

After thinking about all parts of this experiment explain what affects the radius of any orbiting satellite. _____

_____

_____

_____

What affects the radius of a geosynchronous orbiting satellite?

_____

_____

_____

_____

Activity 11-2
Human vs Instrumental
Exploration of Space

Even before the Challenger disaster in 1987, there were many scientists arguing on both sides of this question of space exploration. Some people were saying that exploration of Jupiter and Venus by instrumented satellites gave more information about these distant planets than would have been available by a satellite containing humans. Other people still argue that space exploration and space experimentation by humans in space is vital to the United States.

There are books and magazine articles available on both sides of the argument. Many of the NASA publications that are available to schools have data on both sides of the question. Your teacher will ask you to get information on one or more of the following topics:

- Human Environments in Space
- Limitations of Instruments in Space
- Cost of Launching Instruments
- Time to Reach Distant Planets
- Space Spin-Offs
- Satellite Repair by Humans
- Long Duration Exposure Facility (LDEF)
- Cost of Launching Humans
- Manufacturing in Space
- Living in Zero Gravity
- Landsat
- Space Stations

When you and the other students have finished research on these topics, you will report on the topic in class. While you are doing your research on your assigned topic, you should try to find answers to a number of questions.

Topic Researched _____

Manned or Unmanned _____

Cost of Project, if available _____

Benefits to:

General Public _____

_____

_____

Government _____

_____

_____

Medicine _____

_____

_____

Safety Problems, if any _____

_____

Other Important Information _____

_____

_____

Use the following two pages to take notes as other students make their reports.

Topic _____

Manned or Unmanned _____

Cost of Project, if available _____

Benefits

Safety Problems

Other Important Information

Topic _____

Manned or Unmanned _____

Cost of Project, if available _____

Benefits

Safety Problems

Other Important Information

When you have heard all of the reports, write your answer to the following question.

Should the United States Space Program concentrate on humans in space or instrumented satellites? Why?

Sources of Information:

NASA Facts - Request specific topic
John F. Kennedy Space Center
Florida 32899

Popular Science magazine
Popular Mechanics magazine
Air and Space Magazine Published by Smithsonian
Discover magazine

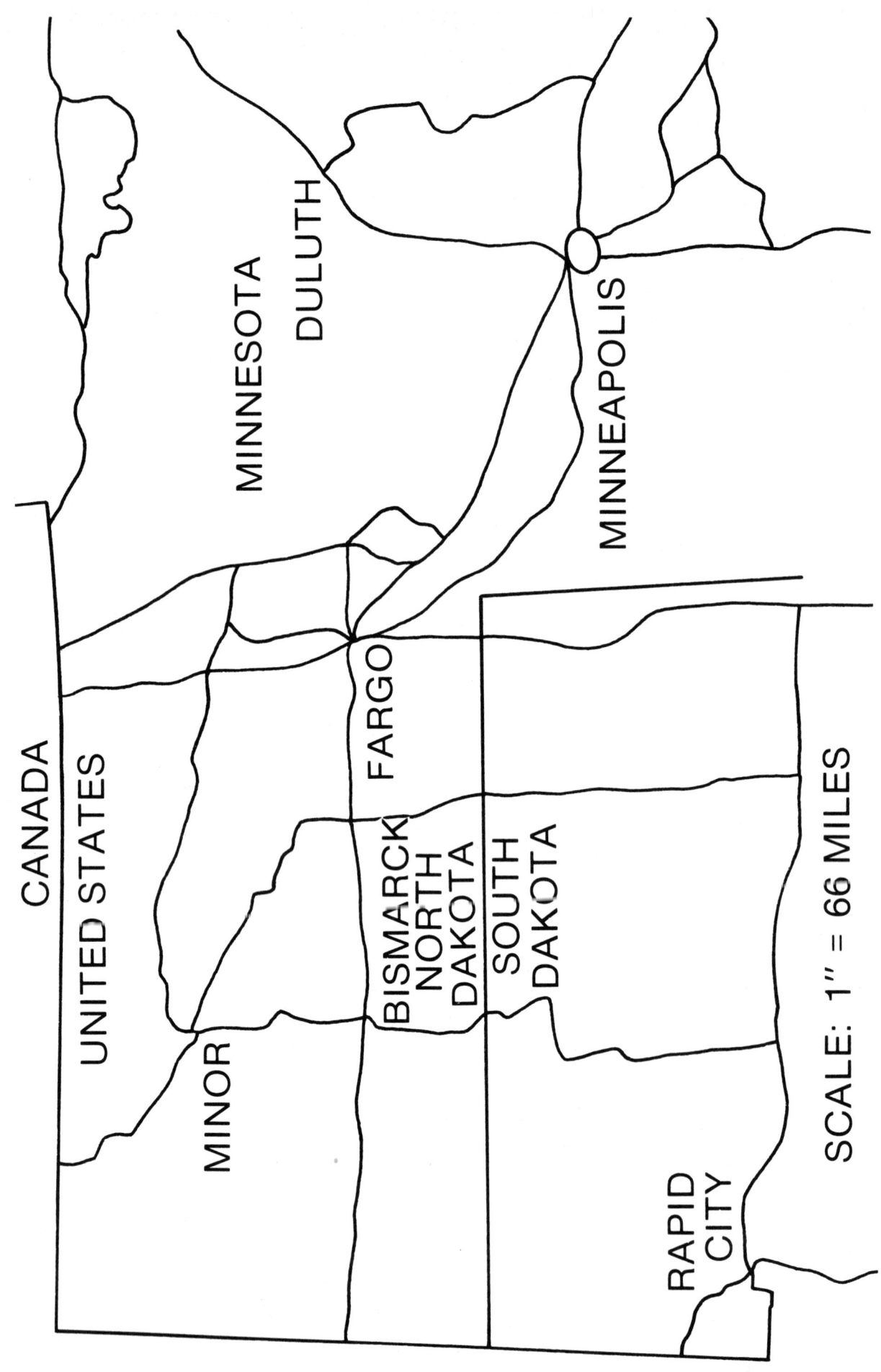

Fig. 11-4 Map for Simulating Looking at Earth from Space

CANADA

UNITED STATES

MINNESOTA

DULUTH

MINNEAPOLIS

FARGO

BISMARCK
NORTH
DAKOTA

SOUTH
DAKOTA

MINOR

RAPID
CITY

SCALE: 1" = 66 MILES

Activity 11-3
See More By Getting Farther Away

One advantage of the use of satellites is that they can "see" a larger portion of earth than can be seen from the surface of the earth or even from a high flying airplane. This ability to "see" more of the world is used in two different ways. Satellites can receive TV signals from one point on the surface and then relay them to another point on earth. Satellites can also take pictures of large areas. Scientists can use these pictures to locate mineral deposits, determine weather patterns, and record crop damage.

In this activity, you will simulate looking at earth from different heights and calculating the area seen from those heights.

MATERIALS

- Map (Figure 11–4 or one from your area)
- Paper, four sheets, 8½" × 11"
- Scissors
- Compass for drawing circles
- Ruler marked in inches

USING THE MATERIALS

1. Draw and cut out a circle, as directed, on each of the four sheets of paper.
2. View the map through the hole in the paper.
3. Using the scale on the map, calculate the area that can be seen from the various altitudes.
4. Using the formula $A = \pi r^2$, calculate the area of the surface that can be seen from the different altitudes.

PROCEDURE

For the map in Figure 11–4, the scale is 1″ = 66 miles. For any other map, you will need to use the scale on that map.

1. A ten-story building is about 200 feet high. To get a picture of how much you can see from 200 feet, draw a circle with ¼″ radius on a sheet of paper. Cut out the circle and place it on the map so that Fargo, North Dakota, is in the center of the circle. Using the scale of the map, calculate the distance you can see, that is, the diameter of the circle in miles, and the area you can see in square miles. Record the information in the table after step 4.

2. Fargo, North Dakota, has the tallest TV tower in the United States (2063 feet). Assume that you are on the top of the tower. Mark and cut a circle ¾″ in radius on the second sheet of paper. Place it on the map so that Fargo is in the center of the circle. The circle shows the area that can be seen from the top of the tower. Calculate the area of this circle and record it.

   This is also the area in which the TV signals could be received from the tower. Only homes in this area can receive programming from that tower. What is the greatest distance from the tower that a home can be to receive that signal? Record this answer.

3. Now measure and cut out a circle with a radius of 1¾″ from the center of the third piece of paper. Again, place Fargo at the center of the circle to show the area that can be seen from an airplane at 10,000 feet. Measure and record the radius and area of the circle.

4. On the last sheet of paper, measure and cut out a circle with a radius of 4″. This simulates what could be seen by a spy plane at 46,500 feet. Place it on the map to make the calculations for the following table.

Altitude	Radius in Inches	Radius in Miles	Area Visible
200 feet			
2,063 feet			
10,000 feet			
46,500 feet			

The size of the circle to be cut was based on the scale of the map and the formula for the distance that can be seen from various altitudes. That formula is $D = \sqrt{1.5h}$ in which D = distance in miles and h is the height above the ground in feet. If possible, use the altitudes given in the activity as "h" and use a calculator that can calculate square root to check your answers for radius in miles by using the formula.

Assume that the total distance that could be seen on the surface of earth is about 12,000 miles. We then need an altitude that gives us a radius of 6,000 miles. Using the distance formula, calculate how high a satellite would need to be in order to see that much of the surface.

$$D = \sqrt{1.5\,h}$$

$$6{,}000 = \sqrt{1.5\,h} \qquad \text{If we square both sides:}$$

$$6{,}000 \times 6{,}000 = 1.5\,h$$

$$\frac{6{,}000 \times 6{,}000}{1.5} = h$$

$$h = \underline{\hphantom{xxxxxxx}} \text{ feet}$$

$$\underline{\hphantom{xxxxxxx}} \text{ feet} \div 5280 = \underline{\hphantom{xxxxxxx}} \text{ miles}$$

The satellites that are used to relay TV signals are in geosynchronous orbit of 26,000 miles above the earth. How does this differ from your calculations? _____

_____

Why is it necessary for the communications satellite to be so far from the surface of the earth?

_____

PROGRAM 12

Risk and Safety with Technology

In the program "Risk and Safety" the point is made that there is no such thing as zero risk. Cavemen and women were at risk from all sort of dangers. People living in the last ten years of the 1900's are at risk from all kinds of dangers. Our risks are different from those of past generations. The people living in the last ten years of the twenty-first century (one hundred years from now) will have risks that are different from ours.

While there is no such thing as zero risk, we do try in many ways to reduce the risk as much as possible.

Activity 12-1
Invasion of Our Living Space

One common risk for human beings throughout our history is the risk of invasion of our living space. The cave, the country, your house, your room, and your car are all at risk of invasion. Present technology allows us to set up systems that will inform us of potential invasion and possibly repel the invader.

In this activity, you will develop a model of a security system. You will need these materials:

- 1 battery
- Switches (3 or more)
- Light bulbs and sockets (2 or more)

Security System A Simple Warning

Using switches and lights, you want to model a simple security system that would warn you when anyone comes up the driveway toward the house. Wire such an alarm system by using one switch as a motion sensor to operate an alarm (a light) inside the house anytime someone came up your driveway.

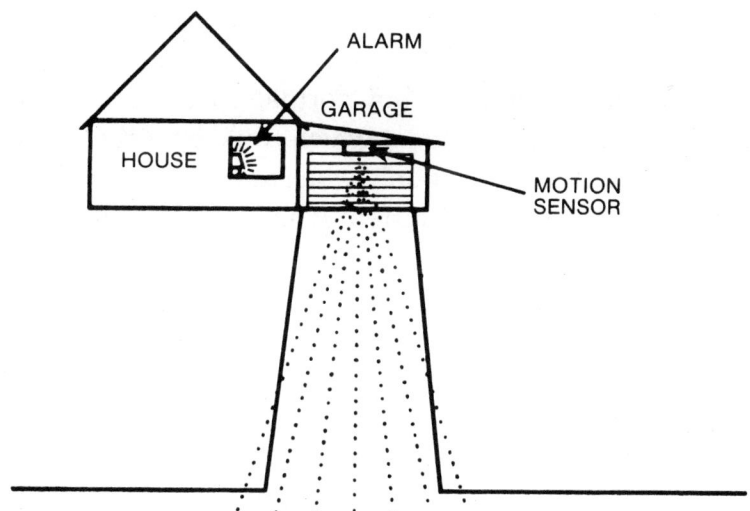

Fig. 12-1 Simple Alarm System

135

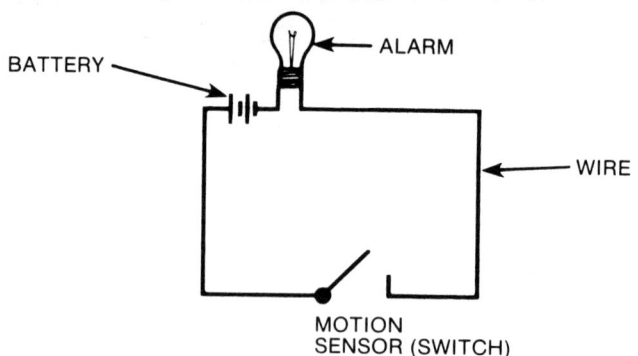

BATTERY

ALARM

WIRE

MOTION
SENSOR (SWITCH)

Fig. 12-2 Wiring of Simple Alarm System

Security System A has some problems. List four of the ways a false alarm may happen:

1. _____

2. _____

3. _____

4. _____

List at least two ways of avoiding alarm detection:

1. _____

2. _____

Security System B

In order to eliminate false alarms caused by normal activities of the mailman or garbage collector, you could install a switch in the house which could be turned off during the day You could also use a bell as the alarm, and set it only when you are away from home.

Draw the wiring diagram for Security System B and construct it.

Security System B (This could also be the model for an automobile alarm system.)

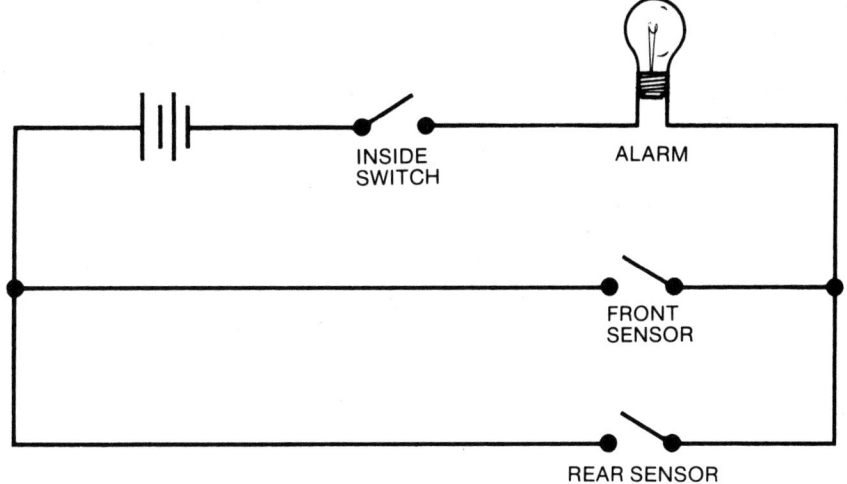

Fig. 12-3   Security System C

Security System C

While Security System B could get rid of some of the false alarms, it did not provide security from anyone coming in from any direction except from the front of the house. Security System C is set up to warn you against anyone coming from either the front or the rear of the house. The wiring for this system is shown in Figure 12–3.

This system provides security from anyone coming in from front or rear. It allows the occupant to turn the system on and off from inside the house.

Security System D

A slight added piece of wiring will allow you to modify Security System C so that if you are not at home an alarm will be activated at the local police station or at a private security center if anyone comes near the house while you are away.

Show the wiring for that system and construct it.

Security System D—Remote Alarm

A problem with this system is that the police or the private security system people will get many false alarms from normal activity such as mail delivery. The sensors must then be altered so that an alarm will sound only when someone tries to enter the house from the front or back. The electrical diagram does not change; only the type of switch changes.

Security System E

A common system now being sold for do-it-yourself installation is made up of a pair of floodlights which go on when someone comes near the house. Obviously you do not want the lights to operate during the day, and you do not want to forget to turn them on each evening at dark.

Draw the diagram for such a system. Label one switch a light sensor and the other a motion sensor. Include a switch inside the house which can be turned off in case you need to get at the wiring of the system. When you finish drawing the wiring for that system, construct the system.

Security System E—Floodlights

Security System F—Optional

Set up the criteria for any other security system you wish to design, such as protecting your desk from invasion; any of four car doors; a missile from space with at least two people required to sound the alarm and fire another missile in return. When the criteria are set, draw the electrical diagram and wire the system.

Activity 12-2
Personal Risk

By your own actions you often put yourself at risk voluntarily. Others may decide on activities which impose a risk on you. Some activities for your benefit impose risks on other people now or in the future. In this activity you will look at some of these activities in terms of the risk involved.

1. Describe an activity in which you take a risk voluntarily.

   _____

   _____

   _____

   _____

   a. Describe the benefits you get from taking that risk.

   _____

   _____

   _____

   _____

   b. Describe how that activity and the risk it involves might affect friends, family, and society.

   _____

   _____

   _____

   _____

c. Describe what you might do to provide the benefits while reducing the risk on you and the effect on others.

_____

_____

_____

_____

2. Describe an activity by someone else that imposes a risk on you.

_____

_____

_____

_____

a. Describe the benefits that you might get from that activity.

_____

_____

_____

_____

b. Describe the benefits that others might get from that activity.

_____

_____

_____

_____

c. Describe how that activity affects others now or later.

_____

_____

_____

_____

d. What recommendations would you make regarding that activity that would provide the benefits but reduce the present and/or future risk?

_____

_____

_____

_____

Compare the activities that you suggested with those suggested by the group.

Activity 12-3
Reducing Highway Risk

Many of today's roads were built 50 to 100 years ago when there were fewer cars. The cars travelled more slowly than they do today. As a result, roads that were considered safe when they were built are hazardous today. Attempts to reduce one hazard often produce another one.

For example, with fewer cars, we had fewer roads and fewer intersections. As the number of intersections increased, traffic was slowed. We built overpasses to get rid of delays. These overpasses were built of concrete and steel. As the road that goes over the overpass gets crowded, we widen it, but we cannot easily widen the road under the concrete overpass. The result is that the right-hand traffic lane is much closer to the concrete than it was at first as shown in Figure 12–4.

The expanded road now has four lanes, each of which is narrower. The right-hand lane is very close to the concrete overpass.

Since there may be five such overpasses in the space of two miles, it costs too much to build new ones. Your job is to figure out a way to reduce the damage to the car and the risk to the passengers in a car if the car runs into the concrete overpass.

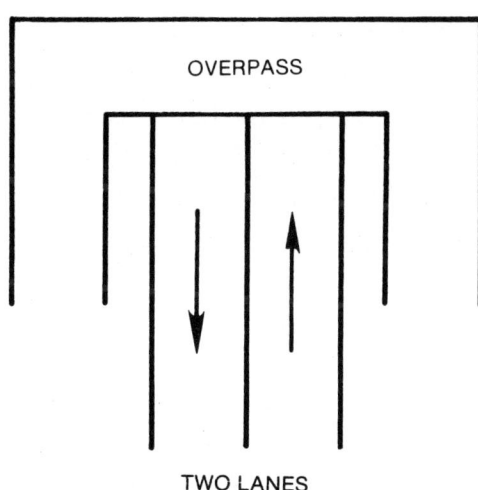 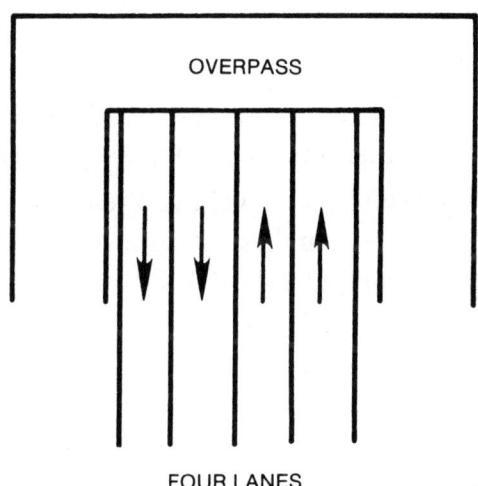

Fig. 12-4   (a) Original Road; (b) Expanded Road

MATERIALS AVAILABLE

- Roller skate car (borrowed from the science lab) or an actual roller skate
- Ramp for the car to travel down to get up speed (Height of the ramp must be adjustable to get different speeds.)
- Board to form a wall for the car to hit
- Modeling clay to make a passenger and front bumper

YOU ARE TO:

Design a system for restraining the passenger to prevent injury.

Design a barrier in front of the wall and/or a car bumper to slow the car enough to reduce the damage to the passengers and the car.

SETTING UP THE EXPERIMENT (AS SHOWN IN FIGURE 12-5)

1. The ramp should be at least 100 cm or about 3 ft in length.

2. Mark lines at 25 cm (15 in), 65 cm (25 in) and 90 cm (35 in).

3. For the slow speed, start the car with the rear wheels at 25 cm; medium speed, 65 cm mark; and fast speed at 90 cm.

4. The wall can be a pile of books, bricks, wood plank, or any other heavy object.

5. The barriers can be anything you wish, but shall not extend more than 7.5 cm in front of the wall. Use your imagination on what those barriers should be, but remember that they must not cost too much and must be easy to replace following a crash.

6. The passenger restraints can be rubber bands, shoelaces, string, balloons, or similar items. Again, use your imagination.

7. The car should have a seat and a dashboard. An electrical outlet box for a wall switch makes a good car enclosure. The brackets can be rearranged to make a seat and a dashboard, as shown in Figure 12-6.

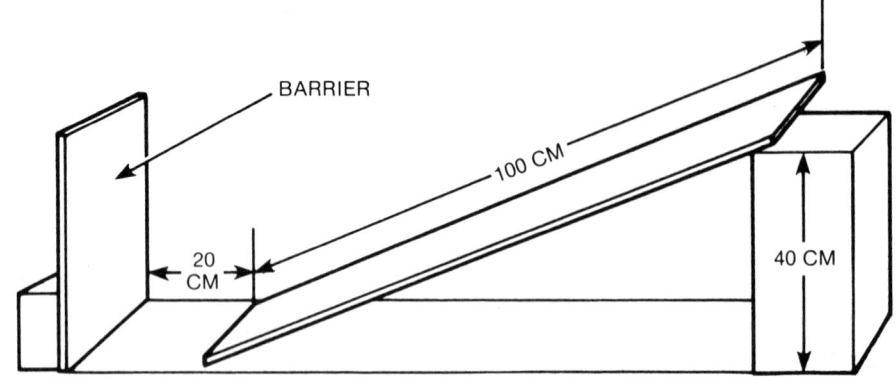

Fig. 12-5 Set-up for Reducing Highway Risk Activity

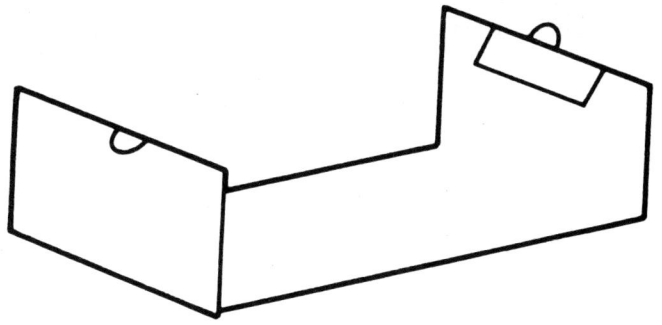

Fig. 12-6 Possible Design for the Car

PROCEDURE

1. Run the car down the ramp at different speeds without any barrier or passenger restraints.

2. In the Passenger Restraints Test table, record any damage to the car and passenger. How much was the bumper compressed or the passenger deformed?

3. Design and install a barrier that can be easily replaced after being hit by a car, and that also reduces the damage to the car.

4. Run the car at the same three speeds after replacing the barrier each time.

5. Record the amount of damage to the car and passenger.

6. Design and install a passenger restraint—a shoulder harness, a seat belt, and/or an air bag.

7. Run the car three times with the barrier and passenger restraint at the same speed as in steps 1 and 4.

8. Record the damage to the car and passenger.

PASSENGER RESTRAINTS TEST						
	Original Thickness of Bumper—Damage Data					
Speed of Car	No Restraint Barrier		Barrier Only		Restraint/Bumper	
	Bumper Thickness	Passenger	Bumper Thickness	Passenger	Bumper Thickness	Passenger
1. Slow—Low Ramp						
2. Medium— Medium Ramp						
3. Fast—High Ramp						

Describe the replaceable barrier you used.

_____

_____

Was it easily replaced? _____

Did it protect the car bumper? _____ the passenger? _____

How did the speed of the car affect the damage?

on the bumper: _____

on the passenger: _____

Describe your passenger restraints. _____

_____

_____

How effective were they? Explain. _____

_____

_____

What recommendations would you make about barriers, passenger restraints, and speed limits as a result of this activity?

Introducing the perfect companions for the **YOU, ME, AND TECHNOLOGY: VIDEO SERIES ACTIVITIES** workbook...

YOU, ME, AND TECHNOLOGY VIDEO PROGRAMS

Agency for Instructional Technology

Delmar and AIT are the exclusive U.S. distributors of twelve, twenty-minute color video programs which cover specific areas of technology. The **You, Me, and Technology** series helps viewers become effective citizens in this complex technological society. The programs can be used individually or as a series, and they are compatible with *any* technology curriculum.

USE THE CARD BELOW TO ORDER YOUR VIDEOS TODAY!

Program (1/2" VHS format)

1. **Living with Technology** (consumerism)/ No. 3417–6
2. **Decisions, Decisions, Decisions** (information processing)/ No. 3418–8
3. **The Technology Spiral** (four technology revolutions)/ No. 3419–2
4. **Energy for Societies** (alternative energy sources)/ No. 3420–6
5. **Health and Technologies** (costs and benefits to society)/ No. 3421–4
6. **Feeding the World** (agricultural technologies)/ No. 3422–2
7. **Communications** (the expanding world)/ No. 3423–0
8. **The Changing Romance** (Americans and wheels)/ No. 3424–9
9. **China, Japan and the West** (transfer of technologies)/ No. 3425–7
10. **Population Patterns and Technology**/ No. 3426–5
11. **Exploring Space**/ No. 3427–3
12. **Risk, Safety, and Technology**/ No. 3428–1

And don't miss...

13. **Teaching with You, Me, and Technology**/ No. 3431–1

This video will show you how to best incorporate the content of the video series into your technology curriculum!

PLACE IN AN ENVELOPE AND MAIL TODAY!

YOU, ME, AND TECHNOLOGY VIDEO PROGRAMS ✂ Order Card

☐ **Yes,** please send me the following programs ($150 each; $1,395 for programs 1-13). 1/2" VHS format

_____   _____
_____   _____
_____   _____
_____   _____
_____   _____
_____   _____

Check one:

☐ My purchase order number* is _____.

☐ A check* is enclosed for $ _____ (include sales tax).

☐ Charge my order to (circle one) [VISA] Visa [MasterCard] Master Card

　　Card # _____ Expiration Date _____

　　Signature _____

*If enclosing check or purchase order, mail to:
delmar publishers inc
2 computer drive, west
box 15015
albany, new york 12212-5015

NAME _____ POSITION/TITLE _____

SCHOOL/INSTITUTION _____

SCHOOL ADDRESS _____

CITY _____ STATE _____ ZIP _____

PHONE NO. (OFFICE) _____ (HOME) _____ HOURS _____

☐ Please have your Delmar representative contact me.　　PRICES SUBJECT TO CHANGE WITHOUT NOTICE.